HIGHLY DISPERSED AEROSOLS

HIGHLY DISPERSED AEROSOLS

by N.A. FUCHS and A.G. SUTUGIN

Translated by
ISRAEL PROGRAM FOR SCIENTIFIC TRANSLATIONS

 ANN ARBOR SCIENCE PUBLISHERS

ANN ARBOR · LONDON · 1970

ANN ARBOR SCIENCE PUBLISHERS, INC.
Drawer No. 1425, 600 S. Wagner Road, Ann Arbor, Michigan 48106

ANN ARBOR SCIENCE PUBLISHERS, LTD.
5 Great Russell Street, London W. C. 1

Library of Congress Catalog Card Number 71–130766
SBN 250 39996 2

This book is a translation from Russian of
VYSOKODISPERSNYE AEROZOLI
Akademiya Nauk SSSR
Moskva, 1969

Table of Contents

The subject matter of this review are aerosols with particle size of less than 1 μ, which differ radically from the conventional coarsely dispersed aerosols in their physical properties, formation conditions, and generation methods. Special methods of physical investigation had to be developed for these aerosols. However, despite the great significance of highly dispersed aerosols in meteorology and a number of industrial branches and despite their basic importance in processes of aerosol condensation in nature and in industry, no review monographs have been published on the subject. It is hoped that the present review will at least partly fill this gap.

INTRODUCTION

1. Fundamental properties of highly dispersed aerosols

Highly dispersed aerosols (HDA) in this review are defined as aerosols with particulate size of less than 1000 Å (by particulate size we shall always mean the diameter d of the particulates). In air at atmospheric pressure, this corresponds to Knudsen numbers greater than unity (the Knudsen number Kn is the ratio of the mean free path of the gas molecules to the particulate radius). The most significant property of HDA with Kn $\gg$ 1 is that the processes of momentum, energy, and mass transfer from the aerosol particulates to the medium and in the reverse direction are described by gas-kinetic equations. In other words, the particulates under these conditions can be simply treated as giant gas molecules. The drag experienced by the particulates moving through the medium, their rate of evaporation (rate of mass loss), rate of heat transfer, and the rate of thermophoresis are thus all proportional to the particulate radius squared. The *aerosol Knudsen number*, i.e., the ratio of the apparent mean free path of the particulates to the particulate radius is also greater than unity for Kn $\gg$ 1, and the rate of coagulation of the HDA is therefore expressed by the standard formula for the collision frequency between gas molecules, where the coagulation constant increases with the increase of particulate size.

For HDA with Knudsen numbers only slightly greater than unity, these formulae should contain correction factors which increase as Kn decreases. For small Kn, i.e., in coarsely dispersed aerosols, all the above processes are governed by entirely different physical laws, which are derived from the equations of hydrodynamics, heat conduction, and diffusion in continuous media.

The vapor pressure of an HDA particulate is substantially higher than the equilibrium vapor pressure above a plane surface of the particulate

3

material. The scattering of light by HDA follows the Rayleigh equation; the scattered intensity is so low that HDA are virtually undetectable by macro-optical methods and in ultramicroscopes HDA particles remain invisible under normal conditions of illumination and observations.

The Brownian movement of the HDA particulates is highly vigorous, whereas their inertia and rate of gravitational sedimentation are negligible, so that diffusion actually governs the settling of HDA on walls and other obstacles. The charges acquired by particulates with radii $\leq 0.03 \, \mu$ in a bipolar ionized atmosphere virtually do not exceed one unit of elementary charge, and the percentage of charged particulates decreases with decreasing diameter.

In all the above respects, HDA markedly differ from aerosols with particulate sizes greater than 1000 Å, which so far have served as the main object of aerosol research. In view of the last remark regarding the electrical charge of HDA particles, their electrical mobility is greater than $10^{-4} \, cm^2/V \, sec$; thus the charged component of the atmospheric HDA belongs to the "medium" and the "heavy" atmospheric ions (to use air physics terminology).

1: FORMATION AND PRODUCTION OF HIGHLY DISPERSED AEROSOLS

2. Formation of condensational HDA

Our discussion of the theory of formation of HDA will start with an analysis of the actual conditions when vapor condensation produces very fine particulates. The mean particulate mass is equal to the ratio of the mass of condensed vapor to the number of particulates formed. Any theory of spontaneous vapor condensation shows that the rate of nucleation (i.e., the rate of formation of nuclei of a new phase) increases faster than a linear function of supersaturation, whereas the rate of condensational growth of particulates is roughly proportional to supersaturation. Therefore, the mean particulate size in an aerosol in the absence of coagulation should decrease with the increase in supersaturation or vapor supercooling (if the temperature is lowered instantaneously). Under real conditions, the cooling rate is finite, and the aerosol formation (including nucleation and condensational growth of the nuclei) continues during the entire cooling phase. For fixed initial and final temperatures, the faster the cooling of the vapor, the smaller is the proportion of the vapor expended in condensational growth of the particulates which formed during earlier cooling stages, at relatively low supersaturation, and the higher is the number of forming particulates.

The steepness of the vapor pressure vs. temperature curve is the main material factor affecting the size of the aerosol particulates. For the same cooling conditions, vapor with a steeper curve will produce aerosols of higher dispersion. For a given material, the steepness of the curve increases with the decrease in temperature; therefore, if a vapor saturated at temperatures T_1 and T_2 ($T_1 > T_2$) is instantaneously cooled by ΔT degrees, a higher supersaturation is reached in the second case and hence a higher degree of aerosol dispersion is achieved. The high supersaturation is not the only relevant factor in this case: the decrease of the equilibrium concentration of the vapor in the forming aerosol also makes a substantial contribution.

7

On the other hand, unless the equilibrium vapor concentration is exceedingly small, the HDA are highly unstable; the small particulates are very rapidly "gobbled up" by the larger particulates. This process is associated with isothermal distillation of vapor from small to large particulates, and also with the effect of local temperature fluctuations in the aerosol: a very slight increase in temperature is sufficient to cause complete evaporation of the small particulates, and the subsequent cooling will readily cause the liberated vapor to condense on the existing larger particulates. Therefore HDA can be produced from materials which are volatile at room temperature only by rapid cooling to the liquid nitrogen point.

HDA generally form by means of spontaneous condensation, since the concentration of impurity condensation nuclei, including gas ions, is small compared to the concentration of nuclei which form spontaneously at high supersaturations. However, when HDA form in plasma burners, or during condensation of cesium vapor in a coronal discharge upon emerging from a de Laval nozzle, and in other cases involving very high concentrations of gas ions, condensation apparently occurs on impurity nuclei.

We have so far ignored the coagulation of particulates, which may involve coalescence or sintering. It will become clear from what follows that coagulation plays a leading role in HDA formation at moderate vapor concentrations.

The classical theory of nucleation [1] has been treated in adequate detail in the literature, and we will only review the fundamental premises of this theory. A supersaturated vapor is regarded as a mixture of isolated molecules and molecular aggregates, whose concentration, according to the general theory of thermodynamic fluctuations, is an exponential function of the free enthalpy of their formation ΔG. The dependence of ΔG on the number of molecules in an aggregate g in a supersaturated vapor is a function with a maximum. There is a certain probability that a given aggregate will grow by accretion of impinging molecules or shrink by evaporation. For aggregates whose size exceeds the critical value g^*, which corresponds to the maximum ΔG, the probability of growth is much greater than the evaporation probability. These "hypercritical" aggregates are considered as nuclei of a new phase. The time variation of the concentration of aggregates containing g molecules is expressed by the equation

$$\Delta f_g/\Delta t = \alpha_{g-1} f_{g-1} - \alpha_g/f_g + \beta_{g+1} f_{g+1} - \beta_g f_g, \qquad (2.1)$$

where α_g is the probability of collision between the aggregate and a vapor molecule, and β_g is the probability of evaporation of a molecule from the aggregate. In the classical theory, β_g is expressed in terms of α_{g-1} on the basis of the principle of detailed balance. When g^* is sufficiently large, the set of equations (2.1) for g from 1 to g^* can be replaced by a differential equation

$$\frac{\partial f(g,t)}{\partial t} = \frac{\partial}{\partial g}\left\{ \alpha_g \frac{\partial \Delta G(g,t)}{\partial g} + \alpha_g \frac{\Delta G(g,t)}{kT} \cdot \frac{\partial \Delta G(g,t)}{\partial g} \right\}, \qquad (2.2)$$

which is essentially an equation of "diffusion" of the aggregate along the continuous g axis. Under steady-state conditions, the rate of nucleation I can be found assuming that the hypercritical particulates are removed from the system and an equivalent mass of vapor is substituted. Under this assumption, we solve the equation

$$-I = \alpha_g \frac{\partial \Delta G(g)}{\partial g} + \alpha_g \frac{\Delta G(g)}{kT} \cdot \frac{\partial \Delta G(g)}{\partial g}. \qquad (2.3)$$

Since HDA formation requires high supersaturations, g^* generally does not exceed 10. Therefore, the classical method of calculating the free enthalpy of the nuclei, whereby the nuclei are considered as liquid droplets which preserve the same properties as any macroscopic volume of the corresponding liquid, obviously does not apply. Moreover, for small g^*, the g axis is discrete and cannot be regarded as a continuum. The process responsible for the establishment of supersaturation. i.e., the cooling of the vapor or the vapor–gas mixture, often proceeds simultaneously with nucleation and particulate growth. Under these conditions, the temperature, vapor concentration, and the supersaturation are all variable, and equations of the form (2.2) cannot be solved.

An alternative approach to the problem of nucleation is known as the *constant-number theory*. Here g^* is assumed to be a small number, only slightly greater than 2, which is almost independent of the nucleation conditions. The probability of evaporation of a nucleus containing $g^* + 1$ molecules is assumed to be negligible, and the concentrations of subcritical aggregates are taken equal to the equilibrium concentrations. Note, however, that in the classical theory the difference between steady-state and equilibrium concentrations of small aggregates is also small. The intuitive considerations which constitute the basis of the constant-number theory were first formulated by La Mer [2] and Courtney [3]. Christiansen [4] proceeded from similar considerations and tried to determine the time to achieve the equilibrium concentrations of the subcritical aggregates after the establishment of supersaturation in a monomolecular vapor; his approach called for a solution of four equations of the form (2.1). These estimates, however, were purely qualitative because of the considerable uncertainty in the value of the free enthalpy of formation of small aggregates.

A practicable method for the calculation of this enthalpy was proposed by Reed [5], who expressed the chemical potential of the inert-gas molecular aggregates in terms of the sums over states of these aggregates. Reed assumed that aggregates of any given size (i.e., with any given number of constituent molecules) exist only in the most stable spatial configurations, that they are subject only to pair interactions between the nearest neighbors, which are described by the Lennard-Jones equation, and that the molecules in the aggregate retain their rotational degrees of freedom. Reed calculated the chemical potentials of aggregates of nitrogen molecules with $g \leqq 8$. The calculations for larger aggregates are more involved and the error associated with the neglect of different spatial configurations becomes considerable. This error can be avoided if Reed's method is replaced by methods which compute the reduced group integrals, but even the latest developments in this direction still require a much greater volume of computations than Reed's original method. Sutugin and Fuchs [6] proposed a method for calculating nucleation at high supersaturations which combined the constant-number theory with Reed's technique. According to this method, the process is subdivided into several tens or hundreds of time intervals Δt. At the beginning of each interval, the concentration of subcritical aggregates is taken equal to the equilibrium concentration at the prevailing temperature and vapor concentration. The number of nuclei of size $g^* + 1$ which form during the time Δt is determined from an equation of the form (2.1), dropping the terms which describe the breakup of the hypercritical nuclei. The vapor consumed in the growth of particles which formed during the earlier Δt is also taken into consideration. This method is applicable to nucleation of associated vapors. In this case, however, one has to consider the probability of collisions between the different associations (if these collisions lead to formation of hypercritical nuclei) and the collisions of associations with hypercritical particles.

The method was applied to calculate the condensation of silver vapor in a hot turbulent argon stream ejected into a cold still air, for silver concentrations ranging between $5 \cdot 10^{14}$ and $5 \cdot 10^{16}$ atom/cm^3 (weight concentration 0.5–50 g/m^3). In these calculations, the time intervals were taken equal to the mean time between consecutive collisions of a critical aggregate with vapor molecules. This assumption simplified the calculations, since all the aggregates of size $g*$ definitely moved on into the hypercritical number region in time Δt. Clearly, Δt did not remain constant during the calculations: it increased as the vapor became progressively depleted. Since Δt was sufficiently small (10^{-5}–10^{-7} sec), the equilibrium concentrations changed at most by a factor of 2 between successive time intervals. On the other hand, Δt was taken large enough for the concentration of the aggregates to assume a new value corresponding to the beginning of the next new time interval. Therefore, in each interval, the concentration of the critical aggregates was regarded as being equal to the equilibrium concentration corresponding to the conditions at the beginning of that interval. The size of the critical nuclei under the above conditions ranged between 2 and 6 atoms, depending on the temperature and concentration, so that the term *constant-number theory* is not completely justified. The concentration of forming particulates was very high, 10^{13}–10^{14} particulates per cm^3, so that coagulation was inseparable from nucleation and condensational growth. Coagulation of very fine particulates may involve coalescence or sintering even at relatively low temperatures. As a result, "continuous" particulates are formed, rather than aggregates. If Δt is sufficiently small, the changes produced during a single time interval by any of the three simultaneous processes—nucleation, condensational growth, and coagulation—are relatively insignificant, and they can therefore be regarded as independent processes following one another in strict succession during every time interval.

Several schemes have been described [7–11] for the calculation of the coagulation of polydisperse aerosols, which involve the solution of a system of linear differential equations of the form

$$\frac{df(g)}{dt} = -\sum_{i=1}^{\infty} K_{ig} f_g f_i + \sum_{j+k=g} K_{jk} f_j f_k. \qquad (2.4)$$

Modern computers are capable of tackling only up to a few hundreds of such equations, and these methods are therefore applicable to processes with at most 100–150 primary particulates at the initial stage. For HDA formation, the primary particulates are the individual molecules, and the smallest value of i in the calculations is 1. For g the smallest starting value is $g^* + 1$; for $g \leqq g^*$ the concentrations are taken equal to the equilibrium values at any given time.

Since the condensation process may produce particulates of 100 Å diameter, i.e., containing 10^5 and more molecules, the above schemes are inapplicable to calculations of moderately dispersed aerosols. Rosinski and Snow [12] proposed a simplified program which assumes that all the particulates are made up of 2^i molecules, where i is an integer. The collision of two particulates of class 2^i produces a particulate of class 2^{i+1}. Collisions of particulates of classes 2^i and 2^{i+1} also produce particulates of class 2^{i+1}, but the yield of these collisions is $(2^i + 2^{i-1})/2^{i+1} = 0.75$ of the number of 2^i particulates which took part in collisions. In collisions between particulates of classes 2^i and 2^{i-j}, where $j > 1$, the size of the 2^i particulates does not change, but their number increases by a factor $(2^i + 2^{i-j})/2^i$. The first of the two assumptions regarding collisions between particulates of different classes slightly overestimates the rate of mass transport along the number axis, whereas the second underestimates this rate, so that the respective errors at least partly cancel out.

In [6], this program was combined with the previous scheme for the calculation of nucleation and condensation growth. Results of calculations and the experimental findings obtained by electron microscopy of particulates sampled at various stages of aerosol formation, reinforced with separate determinations of the sintering properties of the powders obtained by precipitation from the aerosols, led to the conclusion that the process of HDA formation in silver vapor in the relevant concentration range can be divided into six stages.

1. Nucleation and condensational growth.

2. Nucleation, condensational growth, and coagulation, also involving coalescence of particulates.

3. Coagulation with coalescence and condensational growth.

4. Coagulation with coalescence.

5. Coagulation accompanied by coalescence of the fine particulates and sintering of the large particulates.

6. Coagulation producing particulate aggregates held by cohesive forces alone.

The dispersity of the aerosols obtained from measurements of the specific surface of the precipitates was determined almost entirely by coagulation with particle coalescence, so that the mean particulate size was proportional to the cubic root of the vapor concentration.

At very high supersaturations, a labile state may be established, i.e., the thermodynamic barrier between the phases is removed and even two molecules may *irreversibly* form an aggregate. The assumption of a labile state was used by Rosinski and Snow [12] in their calculations of the condensation of Fe_2O_3 vapor and by Stockham [13] for silver vapor. Sutugin and Fuchs, however, established that the labile state is not attained for the condensation of Ag vapor under conditions close to those of Stockham's experiments. Nevertheless, in the final analysis, particulate size is determined by coagulation with particle coalescence, and because of the asymptotic behavior of coagulation, the early stages of condensation hardly affect the final result. This explains the applicability of the Rosinski–Snow scheme to cases when no labile state is established and also shows why Stockham obtained reasonable results.

For $g^* = 2$, the nucleation rate is determined by the relation between the formation rate of a trimer and the dissociation rate of the dimer. The formation of a bound state following a collision of two molecules was considered in [16], but the lifetime of the resulting vapor aggregate, depending on temperature, interaction potential, and the geometrical collision parameters, could be found by numerical methods only. The results obtained in [16] for aerosol formation are therefore inapplicable at this stage. The concentration of the nitrogen dimers calculated by this method is slightly different from the results obtained by Reed's technique [6].

When g^* is sufficiently large and the supersaturation remains fairly constant during the relaxation time of the nucleation process which emerges from the solution of the time-dependent diffusion equation (2.2), we can use the conventional step-by-step method, in which the rate of nucleation for each step is determined from the steady-state solutions of the classical theory. This method is naturally inapplicable in the presence of coagulation or for the condensation of associated vapors. Using the step-by-step technique, Griffin and Sherman [15] have shown that HDA may form by condensation of copper vapor ejected from a de Laval nozzle at high Mach numbers. A detailed description of the step-by-step processes using the steady-state solutions of the classical theory for condensation in nozzles will be found in [16, 17].

When the supersaturation is slowly variable, the condensation can be described by means of differential equations which express the time variation of temperature, vapor concentration, and nucleation rate. In cases of condensation in nozzles, equations of continuity and adiabatic expansion should be added. These equations can be solved analytically for the simplest cases, by introducing numerous simplifying assumptions, and the result is a theoretical dependence of the number of particulates formed on particle mass and time. Solutions of this kind were obtained by various authors [18–20]. Buikov and Bakhanov [21] tried to avoid the simplifying assumptions, and were led to apply numerical computer methods.

The condensational formation of HDA is often accompanied by chemical reactions in the gaseous phase, e.g., the production of carbon black by hydrocarbon pyrolysis, production of metals by decomposition of chlorides and carbonyls, etc. We distinguish between three cases of aerosol condensation involving chemical processes.

1. The chemical reaction which produces the condensing vapor is so fast that it is over before the nucleation stage begins. In this case, the condensation can be treated as a normal "physical" condensation.

2. The chemical reaction proceeds simultaneously with condensation, but the aerosol particulates forming in the process do not affect the chemical reaction. In this case, the chemical reaction should be allowed for in every individual step of the step-by-step calculation of condensation. The factors to be considered are no longer limited to vapor losses in particulate growth and formation of new particulates: the production of additional quantities of vapor by the chemical reaction and, if necessary, the heat effect of the reaction must be introduced into the calculations. A computation program of this kind was described by Dunham [22], who derived a dependence between the reaction rate of H_2SO_4 formation and the rate of nucleation in sulfuric acid vapor.

3. The chemical reaction is catalyzed on the surface of the aerosol particulates. The calculations involve considerable difficulties, and no adequate model of the process has been devised so far. Tesner [23] proposed a method for calculating the time variation of the number of pyrolytic carbon black particles, assuming a given particle size distribution in the final product aerosol.

3. HDA production by physical condensation of vapor

The methods of HDA condensation can be classified according to the technique used to achieve vapor supersaturation: in *physical condensation* the supersaturation is attained by cooling the vapor or the vapor–gas mixture, whereas in *chemical condensation* the gaseous reaction product has a very low vapor pressure at the reaction temperature. The vapor can be cooled by mixing with a cold gas, by heat exchange with a cold object (cooler walls, coolant material, etc.), by adiabatic expansion, or by a combination of any of these methods.

The simplest, but least controllable methods of HDA production by mixing calls for the heating of high boiling point solids or liquids in a turbulent stream of a cold gas. The hot vapor at first mixes with the cold gas near the surface of the heated material, and it is there that the initial stage of nucleation takes place. If the gas is stationary or in cases of laminar flow, the mixing is exceedingly slow and, in accordance with Sec. 2, coarsely dispersed aerosols are mostly formed. Serious difficulties are presented by the possible occurrence of volatile impurities in the heated material, which also form highly dispersed aerosols in the cold gas. In most cases, however, these aerosols can be distinguished from the main aerosol in that their concentration gradually decreases as the volatile impurity evaporates from the sample [24]. If the aerosol concentration remains constant for a long time, we evidently have reached the main aerosol component.

In practical implementations, the material to be evaporated is applied in a thin layer onto a refractory wire heated by an electric current. This is a common method for producing HDA of various salts. A significant shortcoming of this technique is that the aerosol concentration and the particulate size steadily diminish with time. In the case of NaCl, at 450°, the particulate size decreases during 5–7 hrs, dropping to 100–300 Å [25]. This is attributed to the gradual recrystallization of the salt on heating, which lowers the rate of evaporation. To obtain uniform aerosols, the NaCl layer should be preconditioned for a few hours, and the working sample should be changed every day. The HDA produced in this way are fairly polydisperse.

HDA also can be generated by heating pure metal wires to high temperatures in air; this method is often applied to Pt and W. The HDA concentration is fairly constant in time and is adequately reproducible, so that these HDA are often used as condensation nuclei for the production of mono-disperse aerosols. The mechanism of HDA formation from these metals apparently involves the creation of a surface oxide layer and subsequent evaporation of the oxide, which is much more readily volatile than the pure metal. The formation of oxide HDA begins at temperatures when the metal vapor pressure is vanishingly small: at 200°C [28] or 255°C [27] for platinum; 300 or 500°C [28] for tungsten; 650°C [27] and 950°C [24] for nickel. These large discrepancies in the temperatures at which HDA formation begins are attributed to technical difficulties in the detection of aerosol particulates of almost atomic size and to the inevitable presence of impurities in metals. The oxide aerosols are highly polydisperse [27, 29]. When an air current is blown at a rate of 1 liter/min around a platinum wire heated to 300°C, the aerosol concentration reaches about $2 \cdot 10^5$ cm^{-3} and the mean particulate size is about 40 Å [28]. A nichrome wire heated to 1000°C produced a CrO_3 aerosol with particulate size of about 15 Å [29].

A similar method calls for sublimation in a low-pressure gas. The sample is placed in a closed vessel containing a low-pressure inert gas and is heated either by a tungsten coil [30–32] or inductively [33, 34]. Gen, Ziskin, and Petrov [30] were the first to apply this method to prepare aluminum aerosols; later they used it for a wide range of other metals. The particulate size obtained by this method decreases as the rate of evaporation of the metal and the gas pressure decrease. The minimum mean particulate size was around 200 Å. Smaller particulate sizes are unattainable, since at low vapor concentrations the particulate size is virtually independent of the gas pressure and the vapor concentration. Turkevitch [34] studied in detail the mechanism of aerosol formation by this method. He used a device which sampled the aerosol at various distances from the vapor source. An electron microscope enabled him to follow the entire process of aerosol formation or at least the last stages of the process. His conclusion was that the aerosols formed in three stages, corresponding to stages 1, 3, and 5 of the list on p. 12. The decrease of particulate size with the decrease in the inert gas pressure was attributed by Turkevitch to the increase in the rate of diffusion of the vapor, which speeds up the cooling.

A mixer-type variant of this method was applied in [13]: the silver vapor from an inductively heated crucible was allowed to mix with a stream of low-pressure argon flowing past the crucible. Silver aerosols with mean particulate size of at least 40 Å were obtained. The particulate size increased with the increase in vapor concentration, but proved to be virtually independent of the argon pressure. The mixing of the vapor with the cold gas in this case was probably governed by turbulence.

Generation of aerosols in low-pressure gases is clearly a most suitable technique for studying the properties of aerosols, since large changes of pressure in an aerosol involve formidable problems of particle losses, particle coagulation, etc. For this very reason, however, the method is suitable for investigating aerosols under real, normal conditions.

Better HDA generators are those where a relatively large quantity of matter is heated in a thermostated furnace in a slow gas stream; the gas is virtually saturated with the vapor, and is then rapidly mixed with a cold gas. The vapor concentration under these conditions remains constant in time, and this is obviously conducive to preserving a more or less constant concentration and particulate size of the forming aerosol. Two modifications of this generator are used. In one of these, the mixing occurs in a free gas volume, e.g., when the hot turbulent jet of vapor or vapor–gas mixture is ejected into a large volume of cold gas. All the vapor condenses in the gas volume. In the second modification, the mixing occurs in a tee junction, a nozzle with a diffusor, etc., and a substantial part of the vapor condenses on the walls.

A free-mixing generator was used in [6]. The dependence of the aerosol particulate size on the operating conditions of the generator was studied using an aerosol condensing from MoO_3 vapor in a hot air jet ejected with a velocity of 8–580 m/sec into a cold gas [35]. For vapor concentrations ranging between 0.1 and 100 g/m^3 in the jet, aerosols with mean particulate sizes between 70 and 1000 Å were obtained. The particulate size dropped to a certain limit value as the vapor concentration was lowered and the linear flow velocity was increased; for constant flow velocity, the same effect was achieved by reducing the nozzle diameter. It is remarkable that the aerosols forming by condensation in a jet had a higher dispersity than the aerosols forming at the same vapor concentration in a heat exchange generator (see below). The effect of flow velocity and nozzle diameter was understood using the equation of [6] for the rate of mixing in a submerged jet: this equation shows that the mixing rate indeed increases when these parameters are changed as shown. For moderate concentrations (less than 1 g/m^3), the particulate size did not increase markedly with the increase in the weight concentration, but the trend became more pronounced at higher concentrations. This may possibly be attributed to an increased contribution from coagulation with particle sintering.

The formation of HDA in arc or plasma burners follows a similar course. The material is either incorporated into the electrodes, or is injected in powder form into the gas jet. The jet of incandescent gases is cooled by mixing with a cold gas or an atomized coolant. The cooling vapors apparently condense on air ions. According to the results of electron microscopy [34, 36–37], the particulate sizes in the resulting aerosols range from 25 to 1000 Å, with a mean-mass diameter between 200 and 800 Å. Vaporization of metals in an inert gas atmosphere and in a reducing atmosphere gives pure metal aerosols, whereas vaporization in an oxygen-containing medium produces oxide aerosols. Aerosol particulates generated in an arc or a plasma are generally amorphous. Holmgren, Gibson, and Sheer [37] have shown that finer particulates can be attained by forced cooling of the gases leaving the arc. Selover [38] studied the formation of nickel aerosols by cooling an argon plasma containing nickel vapor. Nickel carbonyl was atomized into a jet of argon fed into a plasma burner. Since carbonyl decomposes at these very high temperatures, a chemical reaction of carbonyl decomposition in all probability preceded the physical condensation. At carbonyl concentrations between 0.006 and 0.025 g/liter, the aerosol particulate size varied between 15 and 150 Å. To achieve rapid cooling, the argon jet was sprinkled with a coarsely dispersed aerosol of cold isopropanol. The average particulate size (as measured by the BET method) was 200 Å, but virtually all the particles fell between 30 and 300 Å. Safronov [39] prepared a highly dispersed carbon black aerosol by condensation of the carbon vapor forming when benzene was decomposed in a plasma.

The plasma method will readily convert into the aerosol state even materials with very high boiling points (except those which are thermally unstable). There is nothing to prevent the conversion of large quantities of these materials into aerosol form. Unfortunately, the mechanism of the processes which take place during condensation in cooled plasma jets has not been studied so far.

Let us now consider aerosol generators where vapor and gas are mixed in narrow ducts. A mixer generator for highly dispersed salt aerosols, e.g., NaCl, was described in [40]. It comprises a quartz tube immersed in an electric furnace, with an 8-mm bore tee at one end. The tube is filled with ceramic rings coated with the material to be evaporated, and air is passed at a rate of 0.1–0.5 liter/min through the tube. The air stream is mixed in the tee with a large volume of cold air (5–20 liter/min). To prevent the vapor from condensing on the walls the generator is provided with an auxiliary heating coil. This generator will produce fairly monodisperse NaCl aerosols with mean particulate size of 14–100 Å. The maximum attainable number density of the particulates increases with the increase in particulate size; it ranges between 10^3 and 10^9 cm^{-3}. By adjusting four parameters—the air flow rate through the furnace and the tee and the current in the furnace and the auxiliary heater windings—one alters the mean particulate size and the aerosol concentration between the above limits. The generator is characterized by very stable operation and ensures excellent reproducibility. A generator of this type was successfully applied in [41] for producing radioactive NaCl aerosols.

Reproducible production of monodisperse dioctyl sebacate aerosols with mean particulate size of 30–300 Å is described in [42]. These aerosols were produced in a KUST generator, described in Sec. 6. A NaCl aerosol with a mean particulate size of 34 Å was used as the source of nuclei. The size of the dioctyl sebacate particulates could be readily altered by adjusting the evaporation temperature, the concentration of the condensation nuclei, and the flow rate ratio of the vapor–gas mixture and the cold air.

The vapor–gas mixture also can be cooled by heat exchange with the cold walls. This is the method employed in the Sinclair–La Mer generator of monodisperse aerosols. La Mer [43] has shown that this generator can be applied to produce highly dispersed sulfuric acid aerosols. According to this report, HDA with a particulate size of 20 Å can be prepared.

Matievic and Kerker and co-workers [44–46] modified the La Mer generator and used it to produce aerosols of various inorganic salts—AgCl, NaF, NaCl. These authors studied in detail the effect of the gas flow rate and the evaporator temperature on the particulate size distribution in an aerosol forming by spontaneous condensation and found that the mean particulate size increased with the increase in these parameters. The reason for the increase in particulate size with the increase in flow rate is not quite clear; unfortunately, the detailed design of the modified generator has never been published. Particulate sizes ranging from 10 to 1000 Å were obtained, and the degree of polydispersity of the aerosol proved to be insensitive to the mean particulate size; the rms deviation was 1.15–1.2.

In a number of sources, heat exchange generators were used to obtain NaCl HDA with high mass concentrations (up to 1 g/m^3) for subsequent precipitation in the form of highly dispersed powders [47–49]. Here again the particle size was found to increase with the increase in vapor concentration. Heat exchange prevents attaining high cooling rates, and hence high supersaturations, and generators of this kind are therefore not particularly suitable for the production of aerosols with particulate sizes of less than 100 Å. The formation of MoO_3 aerosols in a hot air current with molybdenum trioxide vapor passed at a rate of 1–28 liter/min through a water-cooled 20-mm bore quartz tube was studied in [35]. At vapor concentrations of 0.002–8 g/m^3, the mean size of the aerosol particulates calculated from the specific surface of the precipitated powders was found to be 130–1000 Å. No pronounced effect of the flow rate on aerosol particulate size was observed.

Adiabatic expansion provides another cooling technique, in addition to heat exchange and mixing with a cold gas. Vapor condensation by adiabatic expansion occurs in the nozzles of supersonic wind tunnels; this is a highly undesirable side effect, and special measures should be taken in order to avoid it (drying the air, heating the nozzle walls, etc.). On the other hand, the formation of HDA, especially with a strong electric charge, in nozzles is of the utmost importance for future development of electrostatic colloidal engines.

For all substances, except helium, the drop of the equilibrium vapor pressure associated with the cooling in adiabatic expansion is faster than the drop of the hydrostatic pressure. Therefore, a sufficiently strong isentropic expansion of a vapor–gas mixture or even of a pure gas may lead to high supersaturations. The supersaturation may increase at a very fast rate, and therefore even water, a highly volatile substance, will form HDA in wind tunnels. The lifetime of these aerosols, however, is very brief, since the particulates evaporate with the increase in temperature in the compression shock. The size of the ice crystallites forming in a de Laval nozzle at M = 5 was estimated in [50, 51] from their inertial properties and the rate of evaporation in the compression shock. The particle size was about 14 Å; all the particles were approximately of the same size, irrespective of the water vapor concentration. The condensation in de Laval nozzles was studied in [57] and [53] with the aim of producing charged particle beams for electrostatic jet engines. Hg_2Cl_2, $HgCl_2$, and $AlCl_3$ vapors are released through a heated nozzle into a vacuum [57]. The particles are collected on a grid, placed for this purpose in the nozzle. This technique produced particles with mean sizes of 100–1000 Å. In [33], cadmium vapor ejected from a nozzle was allowed to condense in the field of a positive coronal discharge, and particles with sizes of about 100 Å were obtained.

Cooling of a vapor–gas mixture by adiabatic expansion, by mixing with a cold gas, or by radiant heat transfer is also observed in cases of aerosol production by the exploding wire method [54–56]. A high-voltage discharge of sufficient power (about 1 kJ for 1 mg of wire weight) is applied to a thin wire held between massive electrodes. The wire is instantaneously transformed into a plasma filament, which produces an aerosol on cooling. In an inert gas atmosphere all metals give a pure metal aerosol, but only noble metals remain unoxidized when exploded in air. In the presence of nitrogen and oxygen, nitride and oxide particulates are formed. The particulate sizes generally range between 50 and 1000 Å. The mean particulate size depends on the wire material and decreases with the increase in the discharge power. Since the instantaneous dilution of the aerosol cloud produces an aerosol composed of individual, unaggregated spheres [54], we conclude that condensational growth, rather than coagulation, plays the leading role in this case. However, the number density in an exploded cloud is high (about 10^{11} cm^{-3}), and coagulation nevertheless takes place, but only in the hot aerosol, so that it does not involve coalescence.

An aerosol generator described in [41] cools the vapor–gas stream by a combination of two processes—adiabatic expansion and mixing with a cold gas. In this device, a hot vapor–gas jet is ejected at a high nozzle velocity into a cold gas. The method produced, without nuclei, fairly monodisperse selenium aerosols with particulate sizes between 600 and 2600 Å.

4. HDA formation involving chemical reactions in the gaseous phase

Thermal and photochemical decomposition of carbonyl vapor, thermal hydrolysis of metal chloride vapors and silicon tetrachloride, and thermal degradation of hydrocarbons are widely used on an industrial scale for the production of highly dispersed powders of metals, oxides, carbon black and silica smoke through the intermediate stage of aerosols. Extensive patent literature is available on all these processes, and it falls outside the scope of our review. We will therefore consider only those publications whose authors concentrated on the mechanism of aerosol formation in the corresponding processes. The dispersity of the aerosol, metals, and metal oxides obtained in these studies were characterized in terms of the specific surface of the dry precipitates, measured by the BET [Brunauer–Emmet–Teller] method.

The photolysis of iron pentacarbonyl vapor in an oxidizing atmosphere has been repeatedly used in the production of aerosols [57–62]. It was established [61] that a single photon splits off one CO molecule from the carbonyl complex, giving a new compound $Fe_2(CO)_9$, which is then oxidized to Fe_2O_3 in the presence of O_2. Kogan [63] applied this reaction to obtain "molecular condensation." The process of Fe_2O_3 formation is described in detail in [64]. As the exposure time of a mixture of nitrogen and iron pentacarbonyl vapor to the light of a mercury lamp was increased from 0.1 to 6 sec, the degree of decomposition of the pentacarbonyl increased from 10 to 100%. After the exposure, the residual carbonyl was absorbed by passing the aerosol through an activated coal layer. This reduced the contribution from the autocatalytic decomposition of carbonyl on the particle surface. For carbonyl concentration of 36.2 g/m^3 and 10–100% decomposition, this method produced HDA with 22–82 Å particles.

The formation of metal oxide aerosols in thermal hydrolysis of metal chloride vapors was studied by a group of French authors [65, 66], who fed oxygen saturated with appropriate vapors into a hydrogen burner. The chloride concentration was 3–12 mg/liter; the temperature of the oxygen–hydrogen flame reached 1600–3000°K. The mean particle size in the solid precipitates was 100–1200 Å; it decreased with the decrease in vapor concentration. The aerosols obtained in a flame with a maximum temperature of 1500°K were finer than at 3000°K but coarser than at 1900°K. At high temperatures, the aerosols are coarser because of particle sintering, whereas at excessively low temperatures, hydrolysis takes place and the nucleation rate is correspondingly lowered. In a large burner flame, coarser particles are formed; this is clear if we remember that the rate of cooling of a submerged jet decreases as its diameter increases.

Zakutinskii and Blyakher [67] studied the thermal hydrolysis of $AlCl_3$ vapor in a nitrogen carrier, which was mixed with water-saturated air. The hydrolysis took place at 400–450°C. Finer aerosols were obtained than by flame hydrolysis: for chloride concentrations of 5–30 mg/liter, the mean particle size was 50–400 Å. The low temperatures used in these experiments apparently ruled out particle sintering and ensured a higher dispersity of the precipitate than in [65, 66].

A review of the work done on carbon black and silica smoke can be found in [68, 69], and HDA formation in radiolysis of gaseous hydrocarbons is described in [70].

Fuchs and Oshman [71] described the preparation of sulfuric acid highly dispersed aerosols by mixing air streams containing sulfur trioxide and water in a tee. The dispersity measurements of these aerosols are described in Sec. 7.

5. Production of HDA by atomizing

When a liquid is atomized, a wide spectrum of particulate sizes is formed, which certainly includes minute droplets, but this crude method cannot be applied to produce HDA without any large particles or with a negligible coarse fraction. Similarly, HDA cannot be obtained by spraying fine powders, since the strong adhesion between the minute solid particles will prevent complete deaggregation. HDA can be produced, however, by atomizing a highly dilute solution of the desired substance and drying the liquid particles to powder. In [72], this method was applied to produce aerosols of various salts with mean particle sizes of 100–1000 Å. Aerosols with mean particle size of 600 Å were obtained by atomizing salt solutions with concentrations of about 0.1 % [73], separating the large droplets by bubbling, and drying the residual aerosol. A modification of the La Mer generator, known as the Rapaport–Weinstock generator, was applied in [74] to produce monodisperse particles of dioctyl phthalate with sizes of at least 400 Å. The mist obtained by mechanical atomizing of a methanolic solution of dioctyl phthalate was passed through a heater, where both components were evaporated. The vapor–gas mixture was then passed through an air cooler; the vapor condensed on the wall or remained in the vapor phase, whereas the dioctyl phthalate vapor condensed both on the walls and on the nonvolatile residues of the evaporated particles, which acted as nuclei. The particle size could be varied by altering the concentration of dioctyl phthalate in solution.

The production of a monodisperse aerosol with particle diameter $r =$ 320 Å by spraying a virus culture was described by Stern in [75].

The current developments associated with electrostatic aerosol engines and energy converters are reflected in a report dealing with the atomizing of liquid cesium in a strong electric field in vacuum, where particulate sizes of $\geq$ 40 Å were obtained [76]. Although these particulates are suspended in a dilute cesium vapor, and not in a gaseous medium, this system can nevertheless be treated as an aerosol of sorts. For comparison, note that electrostatic atomizing of liquids at normal pressure does not give particulates with diameters less than 2000 Å [77].

2: METHODS OF STUDY OF HIGHLY DISPERSED AEROSOLS

6. Investigation of HDA in a suspended state

Because of the differences in the properties of aerosols with particulate size of less than 1000 Å and coarser aerosol systems, entirely different methods have to be applied in the respective experimental studies. We distinguish between two main groups of methods applicable to HDA, methods based on measurements in the suspended state and methods which measure the rate of settling of the particulates or the properties of the dry precipitates.

Since the minimum size of particulates which are visible through an ultramicroscope under normal illumination conditions is around 1000 Å, and the scattering intensity of HDA in accordance with Rayleigh's law rapidly diminishes with the decrease of particulate size, HDA have an exceedingly low optical density in all but very high concentrations (when they swiftly coagulate and turn into coarsely dispersed aerosols). Therefore, to apply optical methods to HDA, the aerosol particulates should first be built up by making a vapor of some other material condense on them. For example, in Aitken and Scholtz adiabatic counters [78], which are widely used by geophysicists for measurements of the concentration of nuclei, the particulates are built up by condensation of water vapor on their surface. The aerosol is injected into a cylindrical chamber filled with moist air. A piston is moved manually to produce expansion, which leads to adiabatic cooling and condensation of the supersaturated vapor. The water droplets produced by condensation of the vapor on the aerosol particulates are collected on a glass plate, where they are counted through a magnifying glass.

The supersaturation attainable in these counters is generally sufficient for developing nonhygroscopic particulates of more than 100 Å in diameter.

27

Exact determination of the concentration of even smaller particulates involves considerable difficulties, irrespective of the increase in supersaturation. The diffusive motion of the particulates makes them precipitate on the counter chamber walls and the sampler walls during the sampling phase and the brief time preceding the expansion. Nolan and Pollak [79–81] designed a photoelectric adiabatic counter, where the fog does not settle to the bottom, and the concentration of particulates is determined from scattering or extinction in the fog using a photoelectric cell. Initially, counters of this type were calibrated using an Aitken counter, so that all the inherent errors of the older method were automatically transferred to the new and more sophisticated technique. An entirely new calibration counter is described in [81, 82]. Here the concentration of the bulked particles is determined by stereomicrophotography in the suspended state, so that some of the shortcomings of the Aitken counter are eliminated, but the main source of error — diffusive losses of fine particulates — remains. The earlier modifications of the photoelectric counters had a poor reproducibility and revealed a tendency to drift with time [83], but these defects were later eliminated [84]. The calibration of the counter depends on temperature and pressure [85, 88]. The optical density of the fog [85] is a function of its number density n, and these quantities are related by the empirical equality $E \approx n^{0.46} M^{0.54}$, where M is the mass of the vapor condensing from unit volume. Other counters using the same principle were described by Verzar [86] and Rich [87].

A numbers of authors have described automatic adiabatic counters with fairly frequent expansion. i.e., 5 expansion cycles per second [89–90]. The time of stay of the aerosol in the chamber and in the communicating channels is thus reduced and the diffusive losses correspondingly diminish. However, the high frequency of expansion cycles inevitably leads to a certain turbulence of the aerosol stream entering the chamber, and this in its turn may increase the losses. To sum up, it seems hardly advisable to use adiabatic counters for measurements of particulates of less than 100 Å in diameter because of the currently unavoidable losses.

In some mixer-type bulking instruments, the supersaturation is created with the aid of diffusion chambers [91–93]. The attainable supersaturations are too low, however, and will develop only particulates with $d > 400$ Å. Moreover, the flow rates of the aerosol in these chambers are very low.

An original mixer-type method for bulking liquid droplets was proposed by La Mer [43, 94–95]. Dibutyl phthalate and sulfuric acid mists were passed over water and toluene, respectively. The droplet composition rapidly reaches equilibrium with the vapor phase, and the particulates grow to a size which can be studied optically. Simple calculations will give the size of the original droplets if the number density is known.

In [96], the fine particulates were bulked by condensation of water vapor in intermixing HDA with a hot vapor–gas mixture.

An exceptionally successful technique of mixer-type particulate bulking was proposed by Kogan and Burnasheva [97]. In their KUST bulking generator (Figure 1), a cold HDA stream is mixed in a conical diffusor with a stream of a hot gas containing vapors of some high boiling point liquid, such as dibutyl phthalate. The two streams are fed coaxially into the diffusor — the hot vapor–gas mixture along the axis and the HDA at the periphery. Turbulent mixing of the two streams produces supersaturation, whose exact magnitude is controlled by adjusting the flow rate ratio, the absolute flow velocities, the evaporation temperature, the width of the annular gap through which the HDA is injected into the diffusor, and by an appropriate choice of the working liquid. The bulking produces a stable monodisperse aerosol with mean droplet size of a few tenths of a micron, which can be observed by ordinary optical methods. The remarkable feature of the KUST generator is that the final size of the bulked particulates is independent of the HDA number density. This is so because most of the vapor condenses on the diffusor walls, and not in the free volume. The condensation on the walls in fact determines the actual supersaturation of the vapor in the instrument, and hence the particulate size. In an improved model of this device, KUST-4 [63], the upper concentration limit at which the bulked particulate size is still independent of the number density was increased to about 10^5 cm^{-3}. Because of this property of the instrument, the HDA concentration can be determined with a nephelometer without any preliminary calibration. Kogan succeeded in developing individual molecules of certain compounds using this instrument [63].

Measurements of light scattering by bulked and unbulked aerosols make it possible to determine the mean particulate size of HDA with particulates of more than 400 Å in diameter, provided the number density is sufficiently high to produce measurable scattering [97].

Particulate losses in KUST were investigated in [98, 47]. A special mixer-type bulking device was built where the HDA was fed axially into

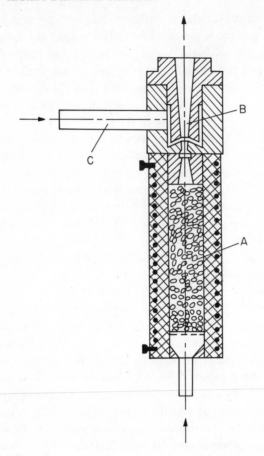

Figure 1

A section through the Kogan–Burnasheva KUST generator.

the diffusor, and not along the periphery as in the original KUST model, and the hot vapor–gas mixture flowed along the walls. The diffusive losses were thus effectively eliminated. A comparison of the bulked aerosol concentrations obtained with the two instruments established that the losses of NaCl particulates measuring 50, 90, and 200 Å in KUST amount to 33, 17, and 10%, respectively. Unfortunately, the modified axial-feed model gave a poorer monodispersity of the aerosols and the particulate size somewhat depended on concentration.

Aitken [99] was the first to measure the activity spectrum of the condensation nuclei, and Nolan [100] observed that the supersaturation required for the development of particulates in his counter could be used to determine their size. Later [101], this method was actually applied to assess the size of NaCl particulates. Low supersaturations can be attained in a so-called "chemical" diffusion chamber, where one of the walls is moistened with water and the other with an HCl solution [101]. This chamber will produce supersaturations between 0.01 % and 1 %, accurate to within 0.01 %. The relative humidity at which the particulates of various inorganic salts of about 200 Å in diameter begin to bulk is close to the result obtained from the standard thermodynamic equations [103, 72]. This method will clearly determine the mean size and even the size distribution in aerosols made up of a single component of known wettability and hygroscopic properties. However, the size distribution can in no way be derived from the activity spectrum of aerosol particulates of unknown composition [104]. The theory of development of aerosol particulates in a diffusion chamber was developed by Buikov [105, 106].

In [107], it was proposed to calculate the size of uncharged aerosol particulates from their gas ion capture coefficient. This coefficient can be found by measuring the decrease in ion concentration when the aerosol is injected into a bipolar ionized atmosphere. However, the number density of the aerosol in this case must be determined independently. As the dependence of the capture coefficient on particulate size is not exactly known, the applicability of this method is dubious.

A flame counter [108, 109] with the aerosol injected into a colorless hydrogen flame was proposed for determining the number density of aerosols which produce a strong glow at high temperatures, e.g., Na or Li salts. The passage of the aerosol particulates through the flame is accompanied by flashes, which are picked up by a photomultiplier. The modification of this instrument described in [110] will record NaCl particulates larger than 400 Å. The upper number density limit of this instrument is limited by the coincidence count errors, and is therefore not very high.

Highly dispersed suspensions have been recently investigated using small-angle scattering of X-rays, which gives the particle size distribution. This method will probably be applied in the near future to more concentrated HDA, e.g., to observe HDA formation under steady-state conditions.

7. Methods involving precipitation of HDA particulates

Of the various methods in this group, we shall first consider those which measure the electrical mobility of the particles, i.e., their speed in an electrical field of 1 V/cm. The classical method of mobility (and concentration) determination of gas ions uses a stream of ionized gas passed through a capacitor; the capacitor current is then measured as a function of voltage. This method was first applied by Langevin [111] to heavy ions, i.e., charged particles of atmospheric aerosols. Since HDA with $d < 600$ Å are virtually free from particulates with more than one unit of electric charge (see p. 72) the mobility spectrum of the aerosol particulates can be used to calculate the size distribution of the charged particulates. Since the percentage of the charged particulates in general increases with the increase in particulate size, this method does not recover the size distribution of all the HDA fractions. Therefore, strictly speaking, the method is applicable only to monodisperse aerosols. Nevertheless, because of its attractive simplicity, it is widely used in measurements of all atmospheric aerosols. In [112], the method was applied to study sulfuric acid HDA. Considerable contribution to the development of this method is due to Israel [113].

A similar method uses a Nolan–Pollak counter with plate electrodes [114]. One measures the concentration of the particulates which remain between the plates after certain fixed time intervals, so that the results give the rate of decrease of the aerosol concentration. The mobility spectrum of the aerosol particulates and the percentage of neutrals are then calculated from the measurement results. Diffusive precipitation of the neutrals on the plates is ignored in these calculations, which may naturally lead to substantial errors for small particulates.

Having determined the percentage of neutrals in a fairly monodisperse HDA in a bipolar ionized atmosphere, we can find the particulate size. To a first approximation, Boltzmann's equation is used. For the fraction of neutrals we have

$$\frac{n_0}{n_0 + n_1 + n_2 + n_3 + \ldots} = \frac{1}{1 + 2\exp\left[\varepsilon^2/2akT\right] + 2\exp\left[(2\varepsilon)^2/2akT\right]},$$

$$(7.1)$$

where n_0 is the concentration of neutrals, n_1 is the concentration of particulates with ± 1 charge units, etc. According to the latest data [115], an exponential analog of the Boltzmann distribution function adequately describes the charge distribution in HDA with particulate sizes of over 200 Å. For aerosols with finer particulates, the percentage of charged particulates under steady-state conditions can only be found approximately (see Sec. 12). This method, originally proposed by Wright [116] and improved by Rich [117], is applicable to particulates of any size, but it cannot be used for polydisperse aerosols. In the latter case, Eq. (7.1) gives a certain equivalent radius a_e. It has been shown [118] that the ratio of a_e to the mean arithmetic radius increases with the increase in the polydispersity of the aerosol, and thus can be used as a measure of polydispersity.

When using methods which measure the steady-state charge distribution, remember that the steady state is attained only after a certain finite interval of time, and under natural ionization conditions it may take as long as several tens of hours [119].

Promising results were obtained with a diffusion method of HDA particulate size measurement, originally used by Townsend to measure the gas ion mobilities. To aerosols this method was first applied by Nolan [120] and Radushkevich [121]. The method measures the diffusive precipitation of particulates on walls from a laminar aerosol flowing through a circular or rectangular cross section channel. The ratio n/n_0 of the aerosol concentrations at the channel inlet and outlet (the *transmission factor*) can be calculated theoretically, and the measurements of this ratio will then give the diffusion coefficient D of the particles, which is related to the mobility B (the velocity of motion under a constant force of 1 dyne) by the equality $D = kTB$ (the dependence of B on particulate size is discussed in Sec. 10). The equations for diffusive precipitation (and for the mathematically equivalent problem of heat transfer in a channel) were derived by various authors in several different forms. In a cylindrical channel, in case of unsteady-state diffusion (when the particulate concentration profile across the channel has a plateau at the level of the initial concentration), the transmission factor is given by the equality [122]

$$n/n_0 = 1 - 2.56\mu^{2/3} + 1.2\mu + 0.177\mu^{4/3} + \ldots \qquad (7.2)$$

where $\mu = Dx/R^2\bar{v}$, x is the channel length, R is the channel radius, $\bar{v}$ is the mean linear flow velocity. For steady-state diffusion ($\mu > 0.005$) [173]

$$n/n_0 = 0.819 \exp(- 3.657\mu) + 0.097 \exp(- 22.3\mu) +$$

$$+ 0.032 \exp(- 5.7\mu). \qquad (7.3)$$

In a plane-parallel channel (with a cross section whose length is much greater than its width $2h$), the concentration ratio for both steady-state and unsteady-state diffusion is adequately expressed by De Marcus's formula [124]

$$n/n_0 = 0.9149 \exp(- 1.885\mu) + 0.0592(- 22.33\mu) +$$

$$+ 0.0258 \exp(- 151.8\mu), \qquad (7.4)$$

where $\mu = Dx/h^2\bar{v}$.

Numerical integration of the equation of diffusion in a plane-parallel channel carried out on a computer [125] gave results which fit De Marcus's analytical solution.

In practice, an aerosol is often passed through a system of parallel channels, a device which Rodebush [126] termed a *diffusion bank*. In this case the expression for μ should contain in the numerator a factor equal to the number of channels. The rate of settling of HDA particulates is so small that there is no need to align the banks vertically for these aerosols. The above expressions are valid for the precipitation of a monodisperse aerosol. In practice, however, all the aerosols are polydisperse, and it is the smallest and the most mobile particulates that are the first to precipitate in each channel. Therefore, the apparent diffusion coefficient of a polydisperse aerosol, which can be calculated from the above equations, increases as the experimental scale parameter $y = \mu/D$ is increased. Hence it follows that diffusion measurements are of any value only if they yield the diffusion coefficient which corresponds to a certain mean particulate size, or else the entire particulate size distribution.

Four methods have been proposed so far for the processing of the results of diffusion measurements,* which give the parameters of the particulate

---

* Recently, Stechkina [163] proposed a fifth method for the determination of the size distribution parameters, which uses the asymptotic expansion of the integrals in (7.6) in terms of the small parameter β_g. Stechkina's analytical expressions can be used to find a and β_g from measured values of D for two different $\bar{v}$.

size distribution. Pollak and Metnieks [127] developed a so-called method of exhaustion, which amounts to the following. If an aerosol is regarded as a mixture of several monodisperse fractions characterized by an ascending sequence of diffusion coefficients $D_1, D_2, \ldots, D_i$ and fractional number densities $p_1, p_2, \ldots, p_i$, then

$$\exp\left(- KD'/\bar{v}\right) = \sum_i p_i \exp\left(- KD_i/\bar{v}\right), \tag{7.5}$$

where D' is the empirical value of the diffusion coefficient, corresponding to a flow velocity $\bar{v}$; K is a coefficient which incorporates the design parameters of the bank according to (7.3) or (7.4). For $\bar{v} \to \infty$, series-expanding the exponential in (7.5), we find $D' = \sum_i p_i D_i$, i.e., $\bar{v} \to \infty$ corresponds to a weighted average D of a mixture of fractions. For $\bar{v} \to 0$, only the term corresponding to the fraction with the smallest D_i need to be retained in (7.5). Therefore, extrapolating experimentally the function $n/n_0 = f(y)$ to $\bar{v} \to 0$, we can find D_1 and p_1. Subtracting p_1 from the total number of particles, we recover the function $f(y)$ for an aerosol without the 1st fraction, and then again extrapolate to $\bar{v} = 0$ to find D_2 and p_2, and so on.

This method became popular with the members of the Irish school [128, 129]. Eq. (7.5) was derived retaining only the first term in the right-hand side of (7.4); if the second term is also retained, the resulting expression is much too cumbersome. Nolan and Scott [130] studied the effect of the number of fractions on the accuracy of the method and found that the aerosol should be subdivided at least into 9 fractions to obtain a correct model of the actual particulate size distribution. However, despite the tedious computation work, this method only gives a moderate accuracy, because of the repeated extrapolations. Fuchs, Stechkina, and Starosel'skii [131] used a computer to calculate the integral diffusion ratio $n/n_0 = f(y)$ for aerosols with a log-normal size distribution $g(a)$ from the equation (see (7.4))

$$n/n_0 = 0.9149 \int_0^\infty g(a) \exp\left[- D(a)y\right] da +$$

$$+ 0.059 \int_0^\infty g(a) \exp\left[- D'(a)y\right] da + \ldots \tag{7.6}$$

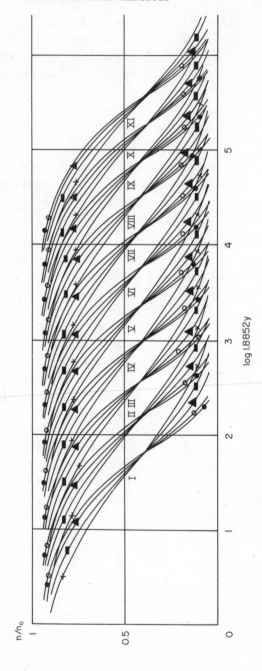

Figure 2

Diffusive precipitation of polydisperse aerosols in a plane-parallel channel:

I $a = 10$ Å; II $a = 16$ Å; III $a = 25$ Å; IV $a = 40$ Å; V $a = 63$ Å; VI $a = 100$ Å; VII $a = 160$ Å; VIII $a = 250$ Å; IX $a = 400$ Å; X $a = 630$ Å; XI $a = 1000$ Å; ● $\log \delta_g = 0$; ○ $\log \delta_g = 0.1$; ▲ $\log \delta_g = 0.2$; ■ $\log \delta_g = 0.3$; × $\log \delta_g = 0.4$.

A family of the $f(y)$ curves were computed, corresponding to a certain mean geometrical particle radius a_g in the range from 10 to 1000 Å and log standard geometrical deviation β_g from 0 to 0.4. The Millikan–Knudsen equation was used for the mobility of the particles (see Sec. 9). Note that the log-normal distribution was observed in aerosols of different origin. Figure 2 plots the $n/n_0 = f(y)$ curve from [131]. In practical applications of this "best fit method", the experimental data are plotted in the coordinates n/n_0 vs. log y and the theoretical curves drawn on a sheet of transparent paper are superimposed on the plot and moved along the horizontal axis until a best fit of the experimental points with one of the theoretical curves is observed. In this way β_g is determined. $\bar{D}$ is obtained from the vertical displacement (along the log y axis) needed to ensure matching with the family corresponding to the given a_g.

The most logical (from the theoretical point of view) approach, which is unfortunately not very practicable, was proposed by Twomey [132]. Transforming Eq. (7.3) or (7.4) for the case of a polydisperse aerosol, he derived an analog of Eq. (7.7) and showed that the experimental dependence $n/n_0 = f(y)$ can be set equal to a convergent series of functions of the form $g(a)A_i e^{-\alpha_i y}$. These are the Laplace transforms of the sought size distribution function $g(a)$, which therefore can be recovered by an inverse Laplace transformation. Theoretically, this method is applicable to aerosols with any unknown size distribution. Twomey's own computations show, however, that if $f(y)$ is calculated from a given $g(a)$, and then $g(a)$ is recovered by this method from $f(y)$, the final result is substantially different from the original $g(a)$. The resulting error soars up because of slight inaccuracies in $f(y)$ associated with measurement errors. It is only when the concentration ratio is determined with an error not exceeding 0.5% over the entire measurement range that this method will produce meaningful results. Since this accuracy is unattainable in routine work, Twomey's method is probably applicable only to estimating the width of the particulate size distribution curve in highly polydisperse aerosols, e.g., in atmospheric condensation nuclei.

Nolan and Scott [130] derived an approximate relation for the determination of the distribution parameters from the diffusion coefficients:

$$D' = \bar{D} - \sigma^2/2K\bar{v}. \tag{7.7}$$

Here $\bar{D}$ is the value corresponding to $\bar{a}$, D' is the experimental value corresponding to the velocity $\bar{v}$, σ is the standard deviation of D. Having determined D for two values of $\bar{v}$, we can find $\bar{D}$ and σ, from which the mean particulate size and the size dispersion are obtained.

Recently, Metnieks [133] carried out a theoretical comparison of three methods for the determination of the size distribution parameters—the method of exhaustion, the best fit method, and the Nolan–Scott method. The method of the inverse Laplace transform [132] was rejected by Metnieks as impracticable. The values of $n/n_0 = f(y)$ obtained by calculations from various size distributions were processed by all three methods. The calculations were carried out for size distributions of the form a^{-1}, a^{-3}, and e^{-a}, as well as for a log-normal distribution and a certain bimodal distribution. The curves obtained in [131] were found to match the function $f(y)$ with all these distributions. The exhaustion method also gave adequate results, whereas the Nolan–Scott method was applicable only to the normal and the log-normal distributions, but even then its accuracy was lower than that of the other two methods. The advantages of the best fit method as compared to the exhaustion method are indicated in Nolan and Scott's study [130].

The solutions of Eqs. (7.2) and (7.4) for diffusion from a laminar flow were obtained assuming a parabolic flow profile along the entire channel. In reality, however, the parabolic profile in a circular pipe is established only at a distance of about $0.05R \cdot \mathrm{Re}$ from the inlet, where Re is the Reynolds number. A detailed analysis shows that diffusive precipitation in the initial section is substantially higher than that predicted by the parabolic flow profile. Since the diffusion parameter at the end of the inlet sections is $\mu_{\mathrm{in}} \approx 0.1D/v$, where v is the kinematic viscosity of the gas, the decrease of the concentration ratio associated with this factor becomes more marked as the diffusion coefficient increases. In practice, this effect is significant only for very fine aerosol particulates, with radii of at most 10 Å. Thomas [134] tried to eliminate the error associated with the inlet effect by measuring the ratio of the transmission factors n/n_0 in two banks 47 and 5 cm long; this ratio was assumed equal to the value of n/n_0 for a bank 42 cm long. This approach, however, may introduce a considerable error in the region of small μ, since n/n_0 can only approximately be taken as an exponential function of length. When the contribution from the second and the third term in Eqs. (7.3) and (7.4) is substantial, the product of the transmission factors n/n_0 for banks of length l_1 and l_2 is not equal to the transmission factor of a bank of length $l_1 + l_2$. The existence of this error was recognized by Thomas himself [135]. Nevertheless, Twomey [136], to obtain points corresponding to various y, proposed passing the aerosol several times through the same bank. This method leads to the same error, since the transmission factor through a bank of length L is different from the square of the transmission factor through a bank of length $L/2$.

An empirical correction for the inlet effect was proposed in [40]. It is applicable to aerosols with a narrow size distribution. For monodisperse aerosols, the values of D calculated for various y from the experimental data should all coincide. However, in reality, D is a function of y even for a monodisperse aerosol; this dependence is attributed to the faster precipitation of particulates in the inlet section compared to the precipitation in the steady-state flow. If y is changed not by altering the flow velocity through the bank, but rather by using banks of different lengths, the relative significance of the inlet effect diminishes as the bank length increases, whereas the absolute magnitude of the effect remains the same for measurements at any y. Therefore, comparison of the concentration ratio n/n_0 from measurements in banks of various lengths will provide an estimate of the inlet effect. We see from Figure 2 that the branches of the calculated curves for $\log \beta_g = 1.0$ and 1.26 coincide for n/n_0. We can thus estimate the inlet effect for aerosols with a certain polydispersity also, provided the comparison is made for $n/n_0 > 0.45$. It was established [40] that this correction is noticeable only for very highly dispersed aerosols; therefore, even for aerosols with $\bar{a} = 7$–10 Å, the correction factor for transmission through the shortest banks (n/n_0 around 0.85–0.9) did not exceed 1.05–1.02.*

The function $f(y)$ is determined experimentally by measuring the concentration of a bulked aerosol before and after the bank. Nolan and Pollak and coworkers used for this purpose an adiabatic counter, and the present authors [40, 42, 98] worked with the KUST flowthrough bulking device [97] (see Figure 1) combined with a photoelectric nephelometer. Sometimes the n/n_0 of radioactive aerosols can be found by radiography of the filters through which the aerosol is passed before and after the bank

---

* Recently, the equation of diffusion of particulates to the walls of a circular channel in the process of establishing the parabolic profile was solved numerically [163]. The incremental precipitation of particulates in the inlet section is 2% for Schmidt number $Sc = v/D = 10$ and 5% for $Sc = 0.9$. This agrees with the empirical correction introduced in [40].

[138, 139]. The measurements of the total filter activity in [139] were compared with the particulate concentrations before and after the bank as measured by an adiabatic counter. The radiographic method gave D values lower by a factor of 2–3 than the adiabatic counter. This is quite understandable since the radioactivity of the activated decay products of the aerosol emanations increases with their size. To determine the diffusion coefficient of the radon decay products, the activity of the residue on the walls of the diffusion channel had to be measured [140]. The results of these measurements were applied to calculate the diffusion coefficient using Berezhnoi's solution [141] of the equation for the precipitation on the walls of products forming directly in the diffusion channel.

No "universal" diffusion bank can be designed for application in the entire range of HDA sizes, since the corresponding diffusion coefficients vary approximately between 10^{-1} and 10^{-5} cm^2/sec. To obtain the required values of μ over the entire spectrum of particulate sizes, the flow velocity through the bank would have to be varied by four orders of magnitude. This is clearly unfeasible, since at exceedingly high velocities the hydraulic drag greatly increases, whereas at low velocities a substantial error is introduced by convection currents. Calculations and the authors' personal experience indicate that for particulates larger than 200 Å, it is advisable to use slit banks, whereas for particles of the order of 10 Å banks of cylindrical channels are better. In [98], for aerosol particulates between 14 and 60 Å, a set of banks of different lengths was used, comprising 8 tubes 6 mm in diameter each, and for particulates between 60 and 200 Å, a set of 60-channel banks made up of tubes 1.4 mm in diameter. In general it is advisable to assemble the banks from a large number of small-bore tubes, rather than from a few wide tubes, since this ensures a wider flexibility of the bank of y values by altering the flow velocity. The use of a set of banks of different lengths for the determination of $f(y)$ eliminates the errors associated with flow velocity measurements, since all measurements are made at the same velocity; it further simplifies the correction for the inlet effect.

The errors which may arise in the applications of the diffusion method, apart from the inlet effect, include the accelerated precipitation of charged particles on channel walls due to electric image forces. Calculations show [142, 143] that this effect is negligible for particulates carrying at most one unit of electric charge, but it becomes significant for multiply charged particulates. A test carried out with a monodisperse aerosol of polystyrene latex by Megaw and Wiffen [144] showed that the diffusion method gives correct results only when grounded metallic banks are used. The latex particles clearly could carry multiple electric charges. In [40] it was established that if the aerosol temperature was higher than the wall temperature, the precipitation for low μ was greater than under isothermal conditions, probably because of thermophoresis and convective transport. In work with very highly dispersed aerosols, constant losses in communication lines can be achieved, although banks of different length are used. To this end, all stopcocks were eliminated in [40], and the change in bank length was offset by moving the KUST instrument on a trolley. The last measure was adopted because, according to the observations of the authors and also according to Nolan and Kenny [145], the precipitation of particulates in bent rubber tubing is much faster than in straight tubes, owing to the electrostatic charge induced when the tubes are bent.

Thomas [135] also made special tests with the diffusion method. He compared the results of optical and diffusion measurements of particulate sizes between 1000 and 3000 Å and came up with a satisfactory fit. The slight discrepancies observed could be readily accounted for by improper correction for the inlet effect. The diffusion method is inapplicable to larger particulates because of the increase in gravitational sedimentation, which cannot be entirely eliminated even by aligning the banks in a vertical position.

The diffusion method can be implemented not only in channels but in principle in any other system of objects which are amenable to theoretical calculations of diffusive precipitation from a flow, e.g., a model of fibrous filters made up of a regular array of parallel cylindrical fibers, perpendicular to the flow direction. The transmission factor of HDA in this model is expressed by the relation

$$n/n_0 = \exp\left(- 2\xi\eta H/\pi R_f\right), \tag{7.8}$$

where ξ is the fraction of the volume occupied by the fibers, H is the thickness of the model, R_f is the fiber radius, η is the capture coefficient of particulates by the fibers, expressed theoretically by [146]

$$\eta = 2.9K^{-1/3}Pe^{-2/3} + 0.624Pe^{-1}, \tag{7.9}$$

where $Pe = 2R_f\bar{V}/D$, $K = \frac{1}{2}\ln \xi + \xi - \xi^2/4 - 0.75$, $\bar{V}$ is the mean flow velocity in the filter. Eq. (7.8) shows a good fit with experimental data [147] on the transmission of HDA with $\bar{a} = 18$–$80\,\text{Å}$ through model filters (Figure 3). The values of the diffusion coefficient in these experiments were determined using diffusion banks. The advantage of this version of the diffusion method is that it is free from inlet effects.

In [148], the diffusion method was combined with measurements of the number of charged particulates; to this end, the diffusion bank was used as a plane-parallel plate capacitor. The measurement of the percentage of charged particulates under steady-state conditions gives the "equivalent" radius of the aerosol particles (see p. 33), from diffusion measurements

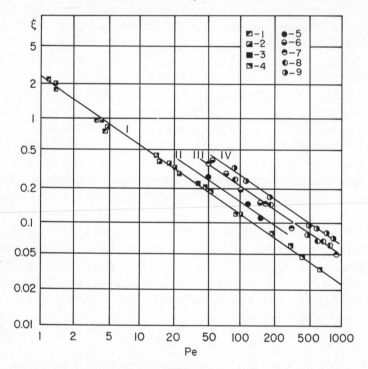

Figure 3

Capture coefficients for diffusive precipitation of particulates in a model filter with parallel staggered cylinders.

theoretical curves: I $\xi = 0.01$;   II $\xi = 0.05$;   III $\xi = 0.135$;   IV $\xi = 0.27$;

mean particle radius, a, Å: 1) 15;   2) 18;   3) 60;   4) 83;   5) 70;   6) 41;   7) 70; 8) 55;   9) 55.

we find some mean particulate radius, and the ratio between these radii provides a qualitative characteristic of polydispersity. The general idea behind this method is quite sound, but unfortunately the values of D measured by the diffusion method depend on y, and the result is thus a function of the aerosol parameters and the experimental conditions.

Some authors [149, 150] used the so-called *static diffusion method*, which measures how the concentration of an aerosol held at rest in a cylindrical or a spherical container varies with time. This method is inapplicable to particulates of about 10 Å, since these particulates precipitate in a few seconds; for larger particulates it is less accurate than the dynamic method because of convection errors.

Labeirie [151] proposed a method of aerosol measurements which calls for radioactivation by a radon impurity. The decay products settle on the aerosol particulates, which are then collected on a filter, the filter is subjected to radiography, and the particulates are counted.

In conclusion, let us consider the methods of HDA study which call for a preliminary precipitation of the aerosol. The actual procedure used in the investigation of highly dispersed precipitates is independent of the particular method used to obtain the precipitate, and we will therefore not go in any detail into such methods as electron microscopy, to which extensive specialized literature has been devoted. We will only consider some specific aspects associated with aerosol precipitates. The main condition to be satisfied is that the sample used for measurements should be sufficiently representative, i.e., the precipitation should in no way be selective. Sampling is often done with thermal precipitators, whose design is described in [152, 153]. The most representative samples are provided by moving-wire thermal precipitators [154]. Electrostatic precipitators are also sometimes used. Since these instruments automatically sort out the aerosol fractions as the aerosol moves along the collecting electrode, measurements should be carried out for samples from different parts of the electrode. Measurements of aerosol precipitates by the BET method, X-ray diffraction, and other methods call for the almost exclusive use of electrostatic precipitators.

An ingenious design of an electrostatic precipitator is described in [155]. A jet of HDA charged in a coronal discharge is injected into a laminar flow between plane-parallel or coaxial electrodes. The fractions are sorted out because the rate of displacement of the particulates toward the collector electrode is proportional to the particulate charge and inversely proportional to the drag on the particulate.

Electron microscopy provides a highly significant method of HDA measurements, since it ensures direct determination of the size and shape of the aerosol particulates and the number density of the HDA. However, the application of electron microscopy to HDA involves certain difficulties, which multiply as the particulate size decreases. These difficulties are associated with the drift and evaporation of the aerosol particulates in the electron beam and the loss of contrast due to the increase of particle transparency. The standard methods of electron microscopy—shadowing of the particles by vacuum sublimation of heavy metals, replica shadowing, the sandwich method [156, 157]—give only limited results, since for very small particulates the thickness of the shadowing layer is comparable to particulate size. It is thus clear why the published photographs are generally those of particulates measuring 100–1000 Å, although numerous authors have actually attained the lower limit of resolution of the electron microscope in their measurements of fine precipitates [32, 44, 45]. In [47], it is indicated that no clear photographs of NaCl particles of less than 200 Å in diameter can be obtained, and in [158] it is shown that Pt and Ni particles of less than 100 Å are either transparent to electrons or evaporate under electron impact. The present authors used an electron microscope with a resolving power of 15 Å and yet failed to resolve the structure of MoO_3 aggregates made up of individual particles of about 100 Å in diameter (this figure was obtained by the BET method). Another open question is the electron microscopy of liquid particulates. If a weighable quantity of the aerosol precipitate can be collected, the mean particulate size is calculated from the specific surface of the precipitate, which is determined by one of the general methods. If the particulates neither sinter nor coalesce in coagulation, but form loose porous aggregates, the specific surface of the precipitate will give the size of the primary, uncoagulated aerosol, but formation of compact aggregates, and especially their sintering, will lead to a certain overestimation of the original particulate size. In [65, 66], the results of the BET method were compared with electron microscopy data, and it was established that both methods give close results for the particle sizes of the precipitates Al_2O_3, TiO_2, ZrO_2, Fe_2O_3 and SiO_2 prepared in the reaction of the vapors of the respective metal chlorides with water vapor at 1600–2000°C.

The results of both methods also match [37] in measurements of the precipitates of metal oxide aerosols generated in an electric arc. However, a decrease of the specific surface of HDA precipitates of NaCl [47, 48] and nickel [38] has been described, which may prove to be a serious source of errors, especially as some time is allowed to pass between the preparation of the precipitate and the measurements.

Membrane filters are also applied to HDA measurements [159–161]. Cartwright [162] proposed a method of replica preparation for electron microscopy which uses particles precipitated on membrane filters. Radiography of filters, membrane filters included, is often applied to radioactive aerosols. Membrane filters are frequently used for the detection of highly dispersed particles with ice-forming activity. The filter, with the particles precipitated on it, is immersed in a supercooled sugar solution, and ice crystals grow on the active particles. The background ice-forming activity of the clean filter should be determined beforehand, since each filter contains 2 or 3 ice nucleating centers.

In conclusion of this chapter, we can say that the existing methods of HDA measurements ensure reliable determination of all the parameters of solid-particle HDA.

3: PROPERTIES OF HIGHLY DISPERSED AEROSOLS

8. Transport processes in HDA. General remarks

The transfer of mass, energy, momentum, and charge from the environment to the particulates plays a leading role in the physics of aerosols. The behavior of the transport processes is determined by the value of the Knudsen number $Kn = l/a$. For $Kn \to 0$, these processes are accurately described by the equations of heat conduction, diffusion, and hydrodynamics of continuous media. For $Kn \to \infty$, the particulates do not distort the Maxwellian velocity distribution of the gas molecules moving toward them, whereas for molecules moving away from the particulates the velocity distribution function can be found without difficulty. Transport processes in this "free-molecular" region are expressed by simple relations which are derived from the molecular-kinetic theory of gases.

The situation is radically different for intermediate values of the Knudsen number: here the theory meets with general mathematical difficulties. The earlier method for solving this problem was by simple "matching": in the immediate vicinity of the particle, the transport processes were assumed to follow the free-molecular mode, whereas at distances of the order of l from the surface of the particulates the transport equations in continuous media were used. Although in certain cases the matching method gave adequate fit with experimental data, it is fundamentally inaccurate. However, this is the only method available to this day for the treatment of charge transport processes.

Exact calculations of the transport rate at intermediate Kn values are possible only by solving the Boltzmann rate equation, and the rate equation cannot be solved unless simplified (linearized). The simplest linearization technique was proposed independently by Wellander and Bhatnagar et al. This method, described in the next section, led to a successful solution of a number of problems in theoretical physics. Brock and Willis applied this method to transport processes in aerosols in combination with a Knudsen iteration, i.e., expansion of the distribution function f into a series in powers of Kn^{-1}:

$$f = f_0 + f_1 Kn^{-1} + \ldots, \tag{8.1}$$

where f_0 is the Maxwellian distribution function. Inserting this expansion in the linearized Boltzmann equation, Brock and Willis proceeded to solve it retaining only the first two terms of the expansion. Their procedure yielded the function f_1, i.e., the first approximation to the distribution function, which they used to calculate the transport rate or the flux of the corresponding quantity toward a spherical particulate. The resulting flux expressions had the form

$$\Phi = \Phi_k/(1 + \lambda Kn^{-1}), \tag{8.2}$$

where Φ_k is the flux in the free-molecular region, λ is some constant. Note that relations of this form are exact only for $Kn^{-1} \ll 1$.

Considerable progress has been recently attained in the theory of transport processes at intermediate Kn values, and the major contributions are all due to authors whose interests lie far from HDA. Sahni, in his work on the theory of neutron transport and absorption in nuclear reactors, carried out an exact calculation of the rate of mass transport to a sphere for any Kn, proceeding from the so-called integral equation. Since the transport processes at high and intermediate Kn values are of great interest for rocket technology, this topic has attracted a considerable number of theoreticians. A variational method developed by Cercignani and Pagani gave the exact heat flux to a spherical particulate and the drag for any Kn. The results of these calculations show excellent fit with the experimental findings.

Studies related to rocket technology are naturally concerned with transport processes at high relative velocities of the particulates and the medium and at large temperature and concentration gradients. In our review, we will only deal with small values of these velocities and gradients, in particular, with velocities which are small compared to the thermal velocity of the molecules.

In conclusion note that the theory of transport to particulates constantly makes use of the concept of the mean free path of gas molecules, a concept which has been entirely eliminated from the modern theory of transport processes in a gas, mainly because its definition involves certain ambiguities. In the theory of transport to particulates, l should be treated not as the mean distance traversed by the molecules between consecutive collisions, but as the mean effective free path l_{eff} which assumes different values for different transport processes. Thus for mass transport l_{eff} is the mean distance that a molecule has moved from the initial position by the time its velocity vector becomes independent of the initial velocity vector. For heat transfer, l_{eff} is the distance at which the molecule acquires the mean temperature of the surrounding gas, for momentum transfer this is the distance at which the velocity acquires the mean (hydrodynamic) velocity of the gas. Without going into a detailed discussion, we may conclude that, because of the incomplete transfer of the energy of the internal degrees of freedom between colliding molecules, l_{eff} for heat transfer should be higher than for momentum transfer. This highly important factor, however, has been allowed for so far only by Lord and Harbour [165].

9. Mass transport in HDA: evaporation and condensational growth of particulates

Evaporation and condensational growth of particulates are the two fundamental processes governing the formation and the development of aerosols. They served as the subject of numerous theoretical and experimental studies, and we will therefore consider these processes in some detail here. For Kn → 0, they are exactly described by the diffusion equation. Generally, a highly idealized process is analyzed: the evaporation (or growth) of an isolated spherical particulate at rest in an infinite volume of an inert (non-interacting) gas. The problem is greatly simplified if the partial vapor pressure is much less than the gas pressure. In this case, we may ignore the collisions between the vapor molecules, and treat the process as quasi-stationary, since the relaxation time of the vapor concentration distribution around the particulate is much less than the time of complete evaporation of the particulate [166]. In what follows, we will only consider this simplified case.

Sufficiently far from the particulate, vapor transport is motivated by diffusion only. The differential equation of the steady-state diffusion of vapor to or from a sphere is written in the form

$$\rho^2 \frac{dn}{d\rho} = I, \tag{9.1}$$

where n is the vapor concentration at a distance ρ from the center of the sphere, I is a constant determined from the boundary condition on the surface of the sphere.

Let us consider two of the simplest cases: (a) evaporation of a particulate in a vapor-free gas, and (b) condensation on a "black" sphere, i.e., a sphere which absorbs all the impinging vapor molecules. Since in these two cases n is proportional respectively to the concentration n_s of the saturated vapor around the particulate or the concentration n_∞ of the vapor for $\rho \to \infty$, integration of (9.1) gives in case a

$$n = - \delta n_s/\rho, \tag{9.2}$$

and in case b

$$n = n_\infty(1 - \delta/\rho), \tag{9.3}$$

where δ is a constant, which is positive for condensation and negative for evaporation. The diffusion flux to the sphere (and from the sphere) is expressed by the relations

$$\Phi = 4\pi D \delta n_\infty \quad \text{and} \quad \Phi = 4\pi D \delta n_s. \tag{9.4}$$

If for evaporation the vapor concentration at infinity is n_∞, and not zero, n in all the following equations should be replaced with $n - n_\infty$ and n_s with $n_s - n_\infty$. A similar substitution should be introduced in case of condensation on a particulate with a non-zero vapor pressure n_s.

All these equations are valid for any Kn, but only at large distances from the sphere. However, for Kn $\to$ 0, vapor transport is a pure diffusive process all the way to the surface of the sphere, the boundary condition for evaporation is $n_{\rho=a} = n_s$, and that for condensation on a black sphere $n_{\rho=a} = 0$. In this case $\delta = \pm a$ and Eq. (9.4) reduces to Maxwell's standard equations

$$\Phi_c = 4\pi D a n_\infty \quad \text{or} \quad \Phi_c = 4\pi D a n_s. \tag{9.5}$$

Note that for Kn $\to$ 0 the rate of mass transport is independent of the evaporation (condensation) coefficient α_c. At any rate, according to a recent report by Jer Ru Mau [167], α_c is close to unity for all liquids.

For Kn $\to \infty$, the rate of evaporation (for $n_\infty = 0$) and condensation (for $n_s = 0$) is expressed by Knudsen's equation for evaporation in vacuum,

$$\Phi_k = \pi a^2 \alpha_c \overline{V}_v n_s \quad \text{or} \quad \Phi_k = \pi a^2 \alpha_c \overline{V}_v n_\infty, \tag{9.6}$$

where $\overline{V}_v$ is the mean thermal velocity of the vapor molecules. The subscripts c and k identify transport in the continuous or the free-molecular region. Note that

$$\Phi_c / \Phi_k = \tfrac{4}{3} \mathrm{Kn} \alpha_c^{-1} \tag{9.7}$$

As we have noted before, the first attempts to solve the problem of particulate evaporation for intermediate Kn values were based on the matching of free-molecular and diffusion fluxes. Some authors [168–170] used matching on the particulate surface, i.e., they assumed that the diffusion equation held all the way to the surface. The resulting expression

$$\Phi = \Phi_c/(1 + 4\mathrm{Kn}/3\alpha_c) = \Phi_k \alpha_c/(1 + 3\mathrm{Kn}^{-1}\alpha_c/4) \tag{9.8}$$

gives a correct functional dependence of Φ on Kn, but the coefficients of Kn and Kn^{-1} are highly exaggerated.

In the "boundary sphere" method [166], the matching is done on the surface of a sphere enclosing the particle, which is distant βl from the particle surface. This is the mean distance that the evaporating vapor molecules will have moved away from the parent particulate when they traverse a distance l (for Kn $\to 0$, $\beta = 1/2$, and for Kn $\to \infty$, $\beta = 1$). This method gives

$$\Phi = \frac{\Phi_c}{\tfrac{4}{3}\mathrm{Kn}/\alpha_c + 1/(1 + \beta\mathrm{Kn})}.$$

For Kn $\ll 1$ and $\alpha_c = 1$,

$$\Phi = \frac{\Phi_c}{1 + \tfrac{2}{3}\mathrm{Kn}}, \tag{9.9}$$

which is very close to the exact expression, but for Kn $\gg 1$ a wrong functional dependence is obtained

$$\Phi = \Phi_k/(1 + 3\mathrm{Kn}^{-2}/4). \tag{9.10}$$

We thus see that the matching methods give unsatisfactory results. Note that Sherman's "universal" interpolation formula [171], proposed by him for a number of transport processes,

$$\Phi = \Phi_c/(1 + \Phi_c/\Phi_k) \equiv \Phi_k/(1 + \Phi_k/\Phi_c), \tag{9.11}$$

coincides with Eq. (9.7) for $\alpha_c = 1$.

An exact solution of the problem of particulate evaporation at intermediate Kn values can be obtained only with the aid of the Boltzmann rate equation. When the vapor concentration is low compared to the gas concentration, as in our treatment, the distribution function of the gas molecules is insignificantly perturbed by collisions with the vapor molecules, and it can be identified to a first approximation with the Maxwellian distribution all the way to the particulate surface. For vapor molecules, on the other hand, the Maxwellian distribution is valid only at distances which are large compared to l. In the case of a "black" sphere, the normal projections of the velocities of all the vapor molecules near the sphere are directed toward the sphere surface, so that the distribution function coincides with "half" the Maxwellian distribution directed toward the sphere. As we move away from the sphere, this pronounced asymmetry in the velocity distribution is gradually smoothed out. Vapor transport at large distances from the sphere is a pure diffusion process, and the only decisive factor is the vapor concentration gradient. As we approach the sphere, the diffusion flux is modified by the flux associated with the asymmetry of the distribution function, and only the latter component remains operative near the surface of the sphere.

The main difficulty in solving the rate equation is connected with the collision integrals, which describe the change in the number of molecules with a given velocity vector in a given volume element owing to collisions with other molecules originating in the same volume. This difficulty stems from the fact that the integrand contains products of distribution functions. Wellander [172] and Bhatnagar et al. [173] suggested replacing the collision integral with simple expressions of the form $v(f_0 - f)d\tau$, where $d\tau$ is the volume element, f is the distribution function in that volume, f_0 is the Maxwellian distribution function, v is the number of molecular collisions in unit time in unit volume (in our case, these are collisions between vapor and gas molecules). This method has been successfully applied to a number of physical problems. Wellander, in particular, used it to calculate the heat transfer from a hot cylinder with a radius comparable to l.

In our problem, this method acquires a clear physical meaning: rather than consider in detail the change in the distribution function of the vapor molecules associated with collisions with gas molecules (the collisions between vapor molecules can be ignored, as we explained before), we assume that the vapor molecules acquire through collisions the isotropic Maxwellian distribution of the gas molecules. This assumption, however, is valid only if the mass ratio of the vapor to gas molecules goes to zero, $m_v/m_g \to 0$. Otherwise, the velocities of vapor molecules show a certain persistence, and their distribution after collisions need not be entirely isotropic.

Brock [174, 175] used this method of linearization of the rate equation combined with the Knudsen iteration in the theory of mass transport for $Kn \geqq 1$. Brock gives only the final expression, without showing the intermediate mathematics. The correction term in his expression contains the mean free path of the gas molecules l_g. Since all the relations in this section contain the mean free path l_v of the vapor molecules, which are contained as an impurity in the gas, we changed Brock's equation using the standard gas-kinetic expressions

$$l_g = (\sqrt{2}\pi n_g d_g^2)^{-1}, \tag{9.12}$$

$$l_v = (\pi n_g d_{gv}^2 \sqrt{1 + m_v/m_g})^{-1}, \tag{9.13}$$

where the subscripts v and g refer to the vapor and the gas, respectively; d is the molecular diameter, and $d_{vg} = 0.5(d_v + d_g)$. Brock's equation thus takes the form

$$\Phi = \Phi_k(1 - \alpha_c 0.807 \sqrt{2}\,\theta Kn^{-1}), \tag{9.14}$$

where θ is a factor related to the persistence of the vapor molecules; it is equal to 0.295, 0.150, and 0.115 respectively for $m_v/m_g = 0$, 1, and 9. These calculations, strictly speaking, are applicable only for $m_v/m_g \to 0$, when Eq. (9.14) can be written in the form (with $\alpha_c = 1$)

$$\Phi = \Phi_k (1 - 0.42\,Kn^{-1}) \approx \Phi_k/(1 + 0.42\,Kn^{-1}) \tag{9.15}$$

We shall see in what follows that for $Kn \gg 1$ this expression gives a fairly accurate result for Φ.

Elementary considerations make it possible to find the lower limit value for the coefficient of Kn^{-1}. Let us calculate the probability that an evaporating molecule traverses a distance l along the normal to the particulate surface, then collides with a gas molecule, and returns to the parent particulate without suffering any additional collisions. This probability is evidently proportional to the ratio of the solid angle subtended by the particulate at the distance l to 4π. Taking into consideration the distribution of the molecular free paths on the way from the particulate and back we obtain for the average probability of a molecule returning

$$\overline{W}_1 = \frac{1}{2l_v} \int\limits_0^\infty \left(1 - \frac{\sqrt{\dfrac{x^2}{a^2} + 2\dfrac{x}{a}}}{2\left(1 + \dfrac{x}{a}\right)} \right) e^{-2x/l_v} dx. \qquad (9.16)$$

For small Kn^{-1}, $\overline{W}_1 = 0.285\, Kn^{-1}$. Since molecules evaporate from the particulate surface at various angles, and not only along the normal, the true result for $\overline{W}_1$ is somewhat higher. The total probability W of a molecule returning to the parent particulate is evidently greater than W_1. Since the rate of evaporation, corrected for the probability of returns, is $\Phi_k = (1 - W)$, the true value of the coefficient of Kn^{-1} is of necessity greater than 0.285.

If the scattering of vapor molecules is anisotropic and their velocities show a definite persistence in collisions, the probability of return after the first collision evidently decreases, and this decrease is more pronounced for stronger persistences, i.e., for higher values of the m_v/m_g ratio. In Equation (9.16), l should therefore be interpreted as the "effective" mean free path of the molecules, l_{eff} (see Sec. 8). A similar interpretation apparently should be applied to Eq. (9.15), as long as there is no exact transport theory which allows for velocity persistence.

Jeans [176] derived the following relation:

$$l_{eff} = l_v(m_v/m_g + 1)/(1 + \beta), \qquad (9.17)$$

where l_v is the mean distance traversed by vapor molecules between collisions; β is a number equal to zero for $m_v/m_g \to 0$, which slowly increases with increasing m_v/m_g and reaches $1/3$ for $m_v/m_g \to \infty$. By (9.17) and 9.13),

$$l_{eff} = (\sqrt{m_v/m_g + 1})/(1 + \beta)\,\pi n_g d_{gv}^2. \tag{9.18}$$

Substituting l_{eff} for l_v in Meyer's equation for the diffusion coefficient of a vapor in a gas

$$D_v = \tfrac{1}{3}\,\overline{V}_v l_v, \tag{9.19}$$

we obtain the Jeans equation [176]

$$D_v = \frac{\overline{V}_v\sqrt{1 + m_v/m_g}}{3\,(1 + \beta)\,\pi n_g d_{vg}^2}, \tag{9.20}$$

which differs from the standard Stefan–Maxwell equation only in the factor $(1 + \beta)$, which is anyhow close to unity. Thus the best policy is to insert in Eq. (9.15) the value of l_v calculated from the coefficient of interdiffusion of vapor and gas using (9.19).

The most exact solution of the problem of mass transport in aerosols for intermediate Kn values was obtained in the theory of neutron transport. One of the fundamental problems of this theory—the so-called Milne spherical problem, which endeavors to calculate the flux of neutrons which reach a black sphere after being scattered isotropically by heavy atoms [177]—is entirely equivalent to our problem for $m_v/m_g \to 0$.

First assume that all the vapor molecules have the same absolute velocity, i.e., the distribution function contains only the vector describing the velocity direction. The collision integrals are thus substantially simplified, and several solutions of the rate equation have been tried. The most comprehensive and exact results were obtained by Sahni [178], who started with the so-called integral equation and applied both analytical and numerical methods in his treatment.

In addition to the exact solution, the "diffusion" approximation is also of interest. In this approximation, the aim is to find a distribution of vapor concentrations around a sphere which satisfies the diffusion equations (9.1)–(9.3) and for large ρ coincides with the exact distributions and therefore gives the exact flux to the sphere from Eq. (9.4). One generally does not calculate the constant δ in this equation, but rather the linearly extrapolated length

$$\lambda = \frac{n_{\mathrm{dif}}(a)}{l_v \left| \dfrac{dn_{\mathrm{dif}}}{d\rho} \right|_{\rho = a}}, \qquad (9.21)$$

whose geometrical meaning is clear from Figure 4, showing the concentration distribution according to the exact solution and according to the diffusion approximation for $Kn = 0.5$ from Sahni's calculations. We see from Figure 4 that at distances $\geq l$ from the surface of the sphere the diffusion approximation gives a concentration distribution which closely fits the exact result, so that the mass transport is governed almost entirely by diffusion; at smaller distances, however, the diffusion approximation is inapplicable. The general results are the same for other Kn values, too.

Comparison of Eqs. (9.3) and (9.21) shows that $\delta = a/(1 + \lambda \, Kn)$, and Eq. (9.4) takes the form

$$\Phi = \frac{4\pi a D n_\infty}{1 + \lambda \, Kn} = \frac{\Phi_c}{1 + \lambda \, Kn}. \qquad (9.22)$$

The values of λ calculated by Sahni are listed in Table 1. The accuracy, according to Sahni's estimates, is 0.3%.

For $Kn \ll 1$, we have from (9.22)

$$\Phi = \Phi_c/(1 + 0.710 \, Kn). \qquad (9.23)$$

For $Kn \gg 1$, λ should be expressed in accordance with the table in the form $\lambda = 1.333 - 0.37 \, Kn^{-1}$ and D should be replaced with $\frac{1}{3}\bar{v}l$. This gives

$$\Phi = \Phi_k/(1 + 0.472 \, Kn^{-1}), \qquad (9.24)$$

which is close to Brock's equation (9.15).

Since tabulated data are not particularly convenient for the calculation of evaporation kinetics, they can be expressed by an interpolation formula

$$\lambda = \frac{1.333 + 0.71 \, Kn^{-1}}{1 + Kn^{-1}}. \qquad (9.25)$$

The maximum deviation of the results calculated using this expression from the figures of Table 1 is $2-6\%$. When calculating the time of evaporation

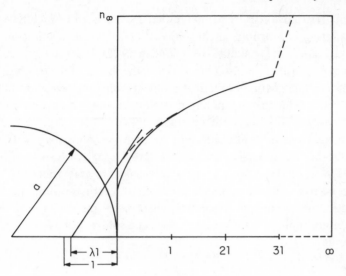

Figure 4

Concentration distribution near an absorbing sphere.

Table 1

THE COEFFICIENT λ ACCORDING
TO SAHNI

Kn^{-1}	λ	Kn^{-1}	λ
0	1.333	1.30	0.997
0.1	1.296	1.50	0.972
0.2	1.353	2.0	0.925
0.5	1.155	2.5	0.892
0.7	1.104	5.0	0.813
1.0	1.043	∞	0.710

of a particulate with the aid of Eq. (9.25), we have to deal with an elementary integral over a rational function of a.

Since the mean absolute velocity of the vapor molecules $\bar{v}_v$ does not appear in the derivation of Eq. (9.24) and in the calculation of λ, the above restriction—equal absolute velocity for all molecules—can be dropped: any velocity distribution may be assumed, including a Maxwellian distribution.

The only restrictive requirements are that $\bar{v}_v$ is conserved in collisions and the scattering is isotropic, i.e., $m_v/m_g \to 0$. Since l is a function of $\bar{v}_v$, the averaged value of l_{eff} should be taken in (9.22).

Velocity persistence is also considered in the theory of neutron transport, but unfortunately Milne's spherical problem with persistence has not been solved yet, and for the lack of any better alternative, we have to introduce the persistence effect as outlined on p. 55.

Since the condensation coefficients α_c for some solids may be less than 1, we also have to consider the case of a "gray" sphere, which absorbs only a fraction α_c of the colliding vapor molecules. A gray sphere is readily seen to be equivalent to a sphere with a finite vapor pressure and $\alpha_c = 1$. Setting n_- and n_+ for the concentrations of vapor molecules near the surface of a gray sphere which move toward the sphere and away from it, we may write

$$n_+ = n_- (1 - \alpha_c) \tag{9.26}$$

The incoming and outgoing molecular fluxes can be expressed in the form

$$\Phi_+ = 2\pi a^2 \bar{v}_v n_- = 2\pi a^2 \bar{v}_v n_+ + 2\pi a^2 \bar{v}_v (n_- - n_+), \tag{9.27}$$

$$\Phi_- = 2\pi a^2 \bar{v}_v n_+. \tag{9.28}$$

These fluxes, however, are also observed for a sphere with a vapor pressure $2n_+$ and $\alpha_v = 1$, when the resultant flux to the sphere is

$$\Phi_{\text{gr}} = 2\pi a^2 \bar{v}_v (n_- - n_+). \tag{9.29}$$

These two spheres are equivalent, i.e., they are characterized by the same vapor concentration distribution in their immediate vicinity. It follows from our previous discussion (see p. 56) that the resultant flux may also be written in the form

$$\Phi'_{\text{gr}} = 4\pi a D (n_\infty - 2n_+)/(1 + \lambda l/a). \tag{9.30}$$

Eliminating n_+ between (9.26) and (9.29) and equating Φ_{gr} and Φ'_{gr}, we obtain an expression for n_- and hence an equation for the flux to a gray sphere,

$$\Phi_{\text{gr}} = \frac{\Phi_c}{1 + \left[\lambda + \dfrac{4(1 - \alpha_c)}{3\alpha_c} \right] \text{Kn}}, \tag{9.31}$$

which was derived by Smirnov by a different method [179].

The effect of capillary forces on evaporation of small particulates is considered in Sec. 15.

The theory should now be compared with the scant experimental material available. Direct measurements of the rate of evaporation of HDA particulates are very difficult to implement, and one is therefore forced to work at low gas pressures (to ensure large *l*). The only remark that can be made concerning the experiments of Bradley et al. [180] and Monchik and Reiss [181] is that these experiments do not contradict the theory, but their accuracy is much too low to settle such unsolved problems as the effect of vapor persistence on the rate of mass transport.

Calculations of the growth and evaporation of aerosol particles under real conditions must take into consideration the heat released by phase transitions, the effect of particulate motion, and other factors. A number of authors considered these aspects. Brock calculated the condensation flux on a moving particulate under free-molecular conditions [182], and in another study [175] he allowed for the heat condensation released in free-molecular growth. He used the Wellander–Bhatnagar model to analyze the unsteady-state growth of a particulate for intermediate Kn values. Smirnov [184] derived an expression for the rate of growth of evaporation of particulates with allowance for the heat of phase transitions and a number of secondary factors. His treatment was based on the results of the theory of neutron transport. Kang Sang-wook [185] applied the method of a boundary sphere to calculate the rate of growth of particulates and also the rate of dissipation of the heat of condensation. The results of his calculations were compared with the experimental data of Weatherstone [186] on the condensation of mercury vapor in a nozzle. This comparison, unfortunately, does not provide a conclusive proof of the validity of the equation derived in [185].

Okuyama and Zung [187] proposed a method for calculating the rate of evaporation of an aggregate of aerosol particulates, and Horn et al. [188] computed the rate of free-molecular evaporation of water droplets injected into a highly rarefied plasma.

In conclusion, let us briefly touch on the topic of gas and vapor adsorption on HDA particulates. The equations for vapor condensation on the particulate surface are clearly also applicable to describe the adsorption kinetics. The only difference is that the accretion probability is a function of the previously sorbed quantity of vapor, i.e., the percentage coverage of the particulate surface with the adsorbate. The total process of vapor adsorption on aerosol particulates can be calculated if the desorption kinetics is known.

In [189], the boundary sphere method was applied to estimate the rate of adsorption of iodine vapor by atmospheric condensation nuclei. The adsorption of iodine was theoretically found to almost reach completion in 30 sec. In practice, however, because of simultaneous desorption, the overall fraction of iodine bound by the particulates does not exceed 75% of the total iodine in the atmosphere, and the equilibrium is reached in no less than 2.5–3 hrs.

The decay products of emanations are almost completely ionized [190], and their capture by aerosol particulates is therefore described by the equations for the rate of charging of HDA particulates by gas ions (see Sec. 12).

10. Momentum transfer in HDA: drag effects

The process of momentum transfer occupies a special position in the theory of transport in aerosols, since this is the only case for which exact and fully reliable experimental data are available (those of Millikan [191] for oil droplets). The theory can therefore be checked in great detail.

For sufficiently small Reynolds numbers, the drag on moving particulates with $Kn \to 0$ is expressed by the Stokes equation

$$F_c = 6\pi\eta a U, \tag{10.1}$$

where U is the particulate velocity, η is the viscosity of the gas.

For $Kn \to \infty$, Epstein [192] derived the equation

$$F_k = \tfrac{4}{3}\pi\beta n_g m_g a^2 \bar{v}_g U, \tag{10.2}$$

where β is a coefficient, whose exact value depends on the mode of reflection of the molecules from the particulate. In case of complete thermal accommodation and diffuse scattering of molecules, $\beta = 1 + \tfrac{1}{8}\pi$; for mirror reflection, $\beta = 1$. Inserting the known expression for the gas viscosity $\eta = 0.499\, n_g m_g v_g l$ in (10.2), we write Epstein's equation in the form

$$F_k = 6\pi a^2 \eta U/(A + Q)l, \tag{10.3}$$

where $A + Q = 1.616$ and 2.250 for the two cases, respectively. By (10.1) and (10.3),

$$F_c/F_k = (A + Q)\,\text{Kn}. \tag{10.4}$$

Millikan fitted the results of his oil drop experiments with an empirical equation

$$F = F_c\left[1 + A\,\frac{l}{a} + Q\,\frac{l}{a}\exp(-ba/l)\right]^{-1}, \tag{10.5}$$

where $A = 1.234$, $Q = 0.414$, and $b = 0.876$. For $l/a = \text{Kn} \to \infty$, equation (10.5) evidently reduces to (10.3). Comparison of the theoretical value of $A + Q$ with the experimental value (which is 1.65) suggests that only a minor fraction of the molecules suffers specular reflection at the surface, whereas the other molecules experience total accommodation and are scattered diffusely.

The first theoretical calculation of the drag for intermediate Kn values was undertaken by Fuchs and Stechkina [193] using the boundary sphere method. Proceeding from not fully justified assumptions, these authors derived for the drag of moving oil droplets

$$F/F_c = \left[\frac{1}{1 + 0.42\,\text{Kn}} + 1.67\,\text{Kn}\right]^{-1}. \tag{10.6}$$

The values of F/F_c calculated from this expression differ at most by 2% from Millikan's data.

Expanding the collision integral in the rate equation in powers of Hermite polynomials and then applying the Knudsen iteration, Liu et al. [194] derived the expression

$$F/F_k = 1 + 0.298\,\text{Kn}^{-1}. \tag{10.7}$$

Linearizing the rate equation as suggested in Sec. 9, Willis [195] used Knudsen iteration to obtain the expression

$$F/F_k = 1 - 0.366\,\text{Kn}^{-1}. \tag{10.8}$$

Cercignani and Pagani [196] applied a new method to solve the transport problem in the range of intermediate Kn values. After linearizing the rate equation, the authors construct a functional of the functions expressing the gas density, the velocity of the molecules, and the temperature; then they apply the variational method to determine the stationary value of the functional. All the above physical functions take on values which correspond to the solutions of the linearized equation, and the stationary value of the functional is used directly to calculate the momentum, heat, and other fluxes. This method led the authors to a successful solution of a wide range of transport problems in systems of different geometry.

Recently Cercignani and Pagani [197] applied this method to determine the drag on spheroidal particulates. The results of their calculations for spherical particulates are plotted in Figure 5, which gives F/F_k as a function of Kn^{-1}. The calculations match the experimental equation (10.5); the slight difference between the two sets of results, never exceeding 2%, is possibly attributable to the use of the value $1 + \frac{1}{8}\pi$ for the coefficient β in the expression for F_k, which presupposes total accommodation of the gas molecules on the particulate surface, whereas in reality a certain fraction of the molecules experience specular reflection. The results of Cercignani and Pagani clearly made a very important contribution to the theory of transport phenomena at intermediate Kn values.

We see from Figure 5 that Willis's equation (10.8) (as could have been expected) gives correct values of F/F_k only up to $Kn^{-1} \approx 0.5$, whereas Eq. (10.7) is definitely wrong. Sherman's "universal" equation (see (9.11))

$$F/F_k = \frac{1}{1 + F_k/F_c},\qquad (10.9)$$

which using (10.4) can be written in the form

$$F/F_k = \frac{1}{1 + Kn^{-1}(A + Q)^{-1}} = \frac{1}{1 + 0.606\, Kn^{-1}},\qquad (10.10)$$

gives somewhat low values of F/F_k (underestimated by as much as 7%). It is remarkable that the values of Φ/Φ_k as a function of Kn^{-1} calculated from Sahni's data for mass transport (Table 1) using Eq. (9.7) closely follow the curve for momentum transfer on Figure 5. At the same time, the curve for heat transfer falls substantially higher.

Brock [198] derived a highly complex expression for the drag on a growing or evaporating particulate at large Kn, which will be of considerable use in calculating the formation of aerosols in rocket nozzles and the effect of the aerosol particulates on the hydrodynamic properties of the gases.

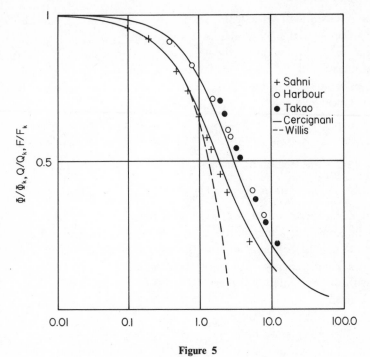

Figure 5

The kinetics of transport processes for intermediate Kn.

11. Heat transfer in HDA: heating and cooling of particulates

The great significance of heat transfer processes between the particulates and the environment is primarily linked with the difference in temperature between an evaporating or growing particulate and the ambient temperature; for particulates with a substantial vapor pressure this factor may greatly affect the kinetics of particulate growth and evaporation.

For Kn → 0, the rate of heat transfer to a particulate is expressed by an analog of Eq. (9.5) derived in the classical theory of heat conduction:

$$Q_c = 4\pi a \chi \Delta T, \qquad (11.1)$$

where χ is the coefficient of heat conduction of the medium, ΔT is the temperature difference between the medium and the particulate. We will only consider the simplest case when ΔT is small compared to the absolute temperature T of the particulate.

To derive an expression for the heat flux to the particulate for $Kn \to \infty$, we introduce as in (9.31) the superscripts $+$ and $-$ identifying molecules which move away from the particle and toward it. The conditions of material balance give the equality $n_g^+ \bar{v}_g^+ = n_g^- \bar{v}_g^-$ or

$$n_g^+ \sqrt{T^+} = n_g^- \sqrt{T^-}. \tag{11.2}$$

The flux of molecules to the particle is

$$\pi a^2 n_g^- \bar{v}_g^- = \pi a^2 n_g^- (8kT^- /\pi m_g)^{1/2},$$

and the mean thermal energy of a single molecule is $c_v T^-$, where $c_v = \frac{3}{2}k$ for monatomic and $\frac{5}{2}k$ for diatomic gases. However, the energy flux to the particulate cannot be found by multiplying these two numbers, since the higher the kinetic energy of the molecules the more frequent are the collisions with the particulate. The exact expression for the energy flux [199] has the form

$$I = \pi a^2 n_g^- \left(\frac{8kT^-}{\pi m_g} \right)^{1/2} (c_v + \tfrac{1}{2}k)\, T^-. \tag{11.3}$$

A similar expression (with the superscripts $-$ and $+$ interchanged) is obtained for the energy flux carried away by molecules, and the resultant net flux is the difference of these two expressions

$$Q_k = \pi a^2 \left(\frac{8k}{\pi m_g} \right)^{1/2} (c_v + \tfrac{1}{2}k) \left[n_g^- (T^-)^{3/2} - n_g^+ (T^+)^{3/2} \right]. \tag{11.4}$$

We introduce the coefficient of temperature accommodation of molecules $\alpha_T = \Delta'T/\Delta T$, where $\Delta'T$ is the difference between the mean temperature of the recoiling molecule and the temperature of the medium; evidently $\Delta'T = T^- - T^+$. Assuming $\Delta'T/T$ to be small, we obtain from (11.4) and (11.2)

$$Q_k = \pi a^2 \left(\frac{8k}{\pi m_g} \right)^{1/2} (c_v + \tfrac{1}{2}k)\, n_g T^{1/2} \Delta'T = \pi a^2 n_g \bar{v}_g (c_v + \tfrac{1}{2}k)\, \alpha_T \Delta T. \tag{11.5}$$

Let us now calculate the ratio Q_c/Q_k. From the standard expression for the thermal conductivity of a gas $\chi = \varepsilon \eta c_v / m_g$, where η is the viscosity of the gas and the coefficient ε is equal to 2.5 for monatomic and 1.9 for diatomic gases, and the expression for the viscosity $\eta = 0.5\, n_g m_g v_g l$, we find

$$\chi = 0.5\, \varepsilon c_v n_g \bar{v}_g l, \tag{11.6}$$

whence

$$Q_c/Q_k = 2\varepsilon c_v \,\text{Kn}/(c_v + \tfrac{1}{2}k)\,\alpha_T. \tag{11.7}$$

Thus, for monatomic gases $\dot{Q}_c/Q_k = 3.75\,\text{Kn}\,\alpha_T^{-1}$, and for diatomic gases $Q_v/Q_k = 3.17\,\text{Kn}\,\alpha_T^{-1}$.

For intermediate Kn, Springer and Tsai [200] used the boundary sphere method to derive the equation

$$Q/Q_c = \left[\frac{1}{1 + \text{Kn}} + B\text{Kn}\alpha_T^{-1} \right]^{-1}, \tag{11.8}$$

where $B = 3.75$ for monatomic and 3.17 for diatomic gases. For $\text{Kn} \to \infty$, this equation takes the form $Q/Q_c = (B\,\text{Kn}\,\alpha_T^{-1})^{-1}$. Using the expression for Q_c/Q_k, we see that in this case $Q/Q_k \to 1$, as is proper.

Using the method of moments, Lees [201] derived for diatomic gases

$$Q/Q_c = (1 + 2.5\,\text{Kn})^{-1}, \tag{11.9}$$

which is definitely inapplicable for large Kn.

Sherman's "universal" formula gives in this case

$$Q/Q_k = \frac{1}{1 + Q_k/Q_c} = \frac{1}{1 + \beta\alpha_T\text{Kn}^{-1}} \tag{11.10}$$

(the value of β should be inferred from (11.7)).

Brock [202] applied the Knudsen iteration to derive the equality

$$Q/Q_k = 1 - \alpha_T B\text{Kn}^{-1}, \tag{11.11}$$

where $B = 0.153$ for monatomic gases and 0.239 for diatomic gases. Expressions of this form, as we have mentioned before, are exact only for large Kn.

The variational method mentioned in the previous section was applied by Cercignani et al. [203] (taking $\alpha_T = 1$) to calculate the values of Q/Q_k for monatomic gases with Kn between 0.0001 and 50. Figure 4 plots the calculated curve.

Until recently, the only experimental determination of the rate of heat transfer to a sphere for low Re numbers and intermediate Kn numbers was that of Takao [204], who measured the cooling kinetics of a heated glass sphere 1.5 cm in diameter, immersed in a vessel about 1 m in dimension, from which the air had been pumped out to the required residual pressure. All the previously mentioned authors compared their private equations (Takao also derived some highly complex expression) with Takao's experimental findings and all came up with excellent agreement. Takao expressed his results in the form of a dependence of Q/Q_c on Kn. In general, Takao's data seem to be sufficiently accurate only for small Kn; the error for Kn $\geqq$ 1 is large, as is evident from the substantial scatter of the experimental points in this range. For Kn > 1, the recent data of Harbour [205] for air are probably more accurate.

Figure 5, in addition to Cercignani's curve, also plots Brock's curve (11.11) (the two curves nearly coincide for large Kn) and the experimental data of Harbour and Takao. Unfortunately, the values of α_T in Takao's and Harbour's experiments were not published, and this complicates the comparison of the theory with experiments.

As we have noted in Sec. 8, heat transfer calculations should use the molecular mean free path characteristic of heat transfer processes l_h, rather than the general free path value l_m for momentum transfer. Lord and Harbour [165] obtained for diatomic molecules $l_h/l_m \approx 1.6$. If Cercignani's curve is recalculated for l_m, and not l_h, it will be shifted to the left so that the abscissae are reduced by a factor of 1.6. Then this curve will virtually coincide with the analogous curves for mass transport and momentum transfer, and we will end up with a single curve for all three processes.

12. Charge transport in HDA: charging and neutralization of particulates by gas ions

Charge transport by the gas ions captured by aerosol particulates constitutes one of the most important physical processes in aerosols. So far, we have no exact idea of what a gas ion is. Numerous pointers seem to indicate that the primary ions form when gas molecules which have lost or gained an electron attract other neutral molecules of constant dipole moment, binding them into an aggregate. The electrical mobilities of the atmospheric "light" ions range between 0.5 and 2.5 $cm^2/V \cdot sec$. The number of molecules in the ion aggregates is therefore variable.

The expressions for the electrical mobility and the diffusion coefficient of ions [199]

$$B = \tfrac{3}{8} \pi^{1/2} l_i \varepsilon \left(\frac{m_g + m_i}{m_g m_i kT} \right)^{1/2}, \tag{12.1}$$

$$D = \tfrac{3}{8} \pi^{1/2} l_i \left[\frac{(m_g + m_i) kT}{m_g m_i} \right]^{1/2}, \tag{12.2}$$

contain the mean free path of ions l_i and the ion mass m_i (m_g is the mass of a neutral gas molecule). We can thus only find the l_i corresponding to a certain m_i. If we take $m_i = m_g$ and use an average mobility value $B = = 1.4 \, cm^2/V \cdot sec$, we find in air at NTP $l_i = 1.3 \cdot 10^{-6}$ cm, and for the mean thermal velocity $v_i = 4.62 \cdot 10^4$ cm/sec.

If, following Junge [206], we take $m_i/A_g = 6$–18, we find $l_i \sim 1.8 \cdot 10^{-6}$ cm and $v_i \sim (1.0$–$2.0) \cdot 10^4$ cm/sec. This uncertainty in the values of l_i and v_i greatly obstructs the comparison with the experimental data on charge transport in aerosols. The process of charge transport to the particulates is determined in this case by the *ionic Knudsen number* Kn_i, which is several times smaller than the molecular number Kn, so that the upper limit of HDA particulate radii, 10^{-5} cm, corresponds to $Kn_i = 0.13$–0.18.

Let us first consider the charging of aerosol particulates by ions of like charge in the absence of an external electrical field. The differential equation for the steady-state diffusion of the ions to a particulate in the electric field set up by the particulate has the form

$$\Phi = 4\pi\rho^2 \left(D \frac{dn}{d\rho} - BEn \right) = \text{const}, \tag{12.3}$$

where E is the particulate field. For $Kn_i \to 0$ we have the boundary conditions $n = n_\infty$ for $\rho \to \infty$ and $n = 0$ for $\rho = a$, so that

$$\Phi_c = \frac{4\pi D n_\infty}{\displaystyle\int_a^\infty \frac{1}{\rho^2} \exp \left[\frac{\phi(\rho)}{kT} \right] d\rho}, \tag{12.4}$$

where

$$\phi(\rho) = \frac{i\varepsilon^2}{\rho} - \frac{\varepsilon^2 a^3}{2\rho^2 (\rho^2 - a^2)} \tag{12.5}$$

is the potential of the electric forces acting on the ion, ε is the elementary charge, i is the number of elementary charges on the particulate (it is positive for ions carrying charge of the same sign as the particulate and negative for ions of unlike charge). The first term in (12.5) is the Coulomb force, and the second term is the image force. The particulate is treated as a conductor of electricity, which is mostly justified for very small particulates. In the derivation of (12.4) it is assumed that the diffusion of ions is a quasi-stationary process. We will later establish under what conditions this assumption is justified.

To determine the significance of the image forces in problems of charge transport in HDA, let us calculate their effect for the diffusion of ions to an uncharged particulate ($i = 0$). The value of the integral in (12.4) in this case is 0.97, 0.83, and 0.62 for $a = 10^{-4}$, 10^{-5}, and 10^{-6} cm, respectively. The image forces in HDA thus should not be ignored.

Taking the case $Kn_i \to \infty$, we note that the mechanism of particulate charging by gas ions involves neutralization of the impinging ions: the charge is transferred to the particulate and neutral molecules recoil. Gentry and Brock [207] not only ignored the image forces but also erroneously assumed a Boltzmann distribution of the ion concentrations around the particulate. Since this distribution is established by molecular collisions, which do not occur near the particulate for $Kn_i \to \infty$, their relation

$$\Phi_k = \pi a^2 \bar{v}_i n_\infty \exp(-i\varepsilon^2/akT) \tag{12.6}$$

is erroneous. Liu, Whitby, and Yu [208] made the same mistake.

A correct solution of the charge transport problem was developed by Keefe, Nolan et al. [209–211], who determined the free path of an ion in the electric field set up by the particulate. Ignoring the image forces, these authors obtained for the flux of ions to an oppositely charged particulate (assuming a Maxwellian distribution of ion velocities)

$$\Phi_k = \pi a^2 \bar{v}_i n_\infty (1 - i\varepsilon^2/akT), \tag{12.7}$$

which coincides with (12.6) only for $i\varepsilon^2/akT \ll 1$. If the image forces are allowed for, the ion flux to a neutral particle is

$$\Phi_k = \pi a^2 \bar{v}_i n_\infty (1 + \sqrt{\pi \varepsilon^2 / 2akT}). \tag{12.8}$$

For charged particles, the problem can only be solved numerically [211]. The values of the "electrostatic" factor θ (the expression in parentheses in (12.7) and (12.8)) are plotted for this case in Figure 6 vs. the particulate radius. The curves in this figure were calculated from the rms ion velocities. Keefe et al. have shown that the curves allowing for the actual ion velocity distribution can be obtained to fair approximation by displacing the curves in the figure to the right, so that the abscissae are multiplied by a factor of 1.5. In their calculations, the authors dealt with ion aggregates made up of 11 water molecules, i.e., they took $m_i/m_g \approx 7$.

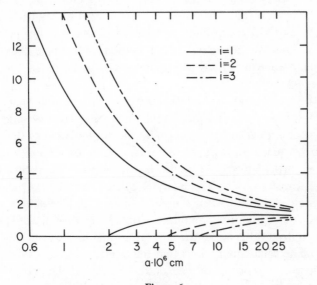

Figure 6

The electrostatic factor for the diffusion of an ion toward a charged particulate.

The problem of particulate charging at intermediate Kn_i involves tremendous difficulties and is still very far from satisfactory solution. Proceeding from Eq. (12.6), Gentry and Brock [207] applied the Knudsen iteration to derive an already familiar expression for Φ/Φ_k. Liu, Whitby, and Yu [208] went even further: they assumed a Boltzmann distribution of ion concentrations near the particulate and a Maxwellian (isotropic) velocity distribution of the ions, the result being that according to their conclusions Eq. (12.6) is applicable at any Knudsen number!

All the other authors solved the problem by the boundary sphere method. The diffusion flux to the sphere was determined using Eq. (12.4), where a has been replaced with the radius of the boundary sphere $a + \beta l_i$. The free-molecular flux to the particulate from the boundary sphere was calculated in different ways. Bricard [212] and Siksna [213] ignored the bending of the ion paths in the electric field of the particulate. Keefe et al. [211] used the above results obtained for $Kn_i \to \infty$ in their calculations, i.e., they assumed that the ions reach the particulate from infinity. Fuchs [214] took the paths to originate at the boundary sphere. Natanson [215] chose the boundary sphere so that an ion colliding inside the sphere could no longer escape outside. The different authors used highly varying values for l_i and $\bar{v}_i$, and any comparison between their results is not particularly meaningful. Since the boundary sphere method in principle does not give exact results, we will not dwell on the relative merits of the various approaches.

We have so far assumed a quasistationary charging and neutralization process. However, every charging event involves a definite change in the distribution of ion concentrations around the particulate. For the charging process to be quasistationary, the relaxation time of the ion atmosphere around the particulate should be much less than the mean time between the charge transfer events. The analysis carried out in [214] shows that this condition is satisfied when $a^3 n_\infty \ll 1$ for $a \geqq 10^{-5}$. For $a = 10^{-6}$ cm, a more exacting condition should be satisfied, $a^3 n_\infty \ll 0.01$. These conditions are indeed observed in the laboratory and in the atmosphere.

Approximate calculations show that the electric field set up by the non-uniform distribution of the ion concentration around the particulates can be ignored under these conditions, i.e., the screening effect is not considered.

If the charge transfer occurs in a bipolar ionized medium, the ions of both charges diffuse independently to the particulate. For any initial charge distribution of the particulates, a steady-state distribution is eventually established in the aerosol, i.e., the number of particulates with i charges of a certain sign which gained one more charge of the same sign in unit time is equal to the number of particulates with $i + 1$ charges which acquired in unit time one charge of the opposite sign. For simplicity, we assume a monodisperse aerosol, in which the concentrations, mobilities, and other properties of the ions of both signs coincide. From the above expressions we can determine the fluxes of like and unlike ions to particulates with i charges, Φ_i and Φ_i', and then write the equations

$$2\Phi_0 N_0 = \Phi_1' N_1; \; \Phi_1 N_1 = \Phi_2' N_2; \dots; \Phi_i N_i = \Phi_{i+1}' N_{i+1}, \quad (12.9)$$

where N_i is the steady-state concentration of particulates with i charges. From these equations we find the charge distribution of the particulates (the coefficient 2 in the first equation is attributed to the fact that an uncharged particulate may capture an ion of any sign). Note that this distribution is independent of the ion concentration, but the rate of establishment of the distribution is proportional to n_∞.

For large particulates ($a \gtrsim 10^{-4}$ cm), when Eq. (12.4) is applicable and the image forces can be ignored, this equation takes one of the following forms:

$$\Phi_0 = 4\pi Dan_\infty; \; \Phi_i = \frac{4\pi Dn_\infty \lambda_i}{\exp \lambda_i - 1}; \; \Phi_i' = \frac{4\pi Dan_\infty \lambda_i}{1 - \exp(-\lambda_i)}, \quad (12.10)$$

where $\lambda_i = |i\varepsilon^2/akT|$. By (12.10), $\Phi_i/\Phi_i' = \exp(-\lambda_i)$.

Multiplying the equalities in (12.9), we get

$$N_i = N_0 \frac{\exp \lambda_i - 1}{\lambda_i} \exp\left(-\sum_1^i \lambda_k\right) = \quad (12.11)$$

$$= N_0 \exp(-\varepsilon^2 i^2/2akT) [\exp(\lambda_i/2) - \exp(-\lambda_i/2)]/\lambda_i.$$

Calculations show that the factor $[\exp(\lambda_i/2) - \exp(-\lambda_i/2)]/\lambda_i$ for particulates with $a \gtrsim 10^{-4}$ cm and moderate charges (the great majority of particulates) is very close to unity. Dropping this factor, we obtain

$$N_i = N_0 \exp(-i^2 \varepsilon^2/2akT), \quad (12.12)$$

i.e., the charge distribution of the particulates is expressed by the Boltzmann equation.

Eq. (12.12) is highly accurate for $a \gtrsim 10^{-4}$ cm, but it becomes less reliable for smaller particulates. In connection with the method described in Sec. 4 for the determination of particulate size from the percentage of charged particulates under steady-state conditions, note that for $a = 10^{-5}$ cm this percentage calculated from the Boltzmann equation is lower by about

2% (rel.) than a more exact figure, for $a = 3 \cdot 10^{-6}$ cm it is lower by about 10%, and for $a = 10^{-6}$ cm it is about one half of the more exact figure.

Keefe et al. [211] assumed a Boltzmann charge distribution a priori, and various explanations were devised to account for the deviations from this distribution at small a. And yet, the Boltzmann equation is applicable only to equilibrium states; if the ions captured by the particulates were to separate as ions (i.e., a system with "detailed balance"), the Boltzmann equation would certainly apply, but in reality we are dealing with steady-state conditions, and not with a detailed balance.

We have already mentioned in the introduction that the number of elementary charges acquired by HDA particulates in a bipolar ion atmosphere is generally small. The table below gives the charge distribution of HDA particulates under steady-state conditions as calculated by Fuchs [214] (rounded off to hundredths).

Table 2

STEADY-STATE CHARGE DISTRIBUTION OF HDA PARTICLES IN
A BIPOLAR IONIZED ATMOSPHERE

a (cm)	10^{-7}	$3 \cdot 10^{-7}$	10^{-6}	$3 \cdot 10^{-6}$	10^{-5}
Fraction of uncharged particles	0.99	0.95	0.76	0.51	0.29
Fraction of particles with 1ε	0.01	0.05	0.24	0.45	0.44
2ε	—	—	—	0.04	0.20
3ε	—	—	—	—	0.06
4ε	—	—	—	—	0.01

In the presence of an external electric field, charging by the current of ions moving in the electric field is superimposed on the regular diffusional charging of the particulates. In case of unipolar ionization, the additional ion flux is expressed by Pauthenier's equation

$$\Phi_E = 3\pi E n_\infty B a^2 (1 - i\varepsilon/3Ea^2)^2, \qquad (12.13)$$

where E is the field strength. To compare this ion "drift" flux with the diffusion flux, let us consider the charging of a neutral particulate with $a = 10^{-5}$ cm. The diffusion flux can be expressed with fair accuracy by Eq. (12.10) for Φ_0. From this equation and from Eq. (12.13) we readily find that

$$\frac{\Phi_E}{\Phi_D} = \frac{0.75\varepsilon Ea}{kT} = 0.9 \cdot 10^4 Ea \qquad (12.14)$$

or, for $a = 10^{-5}$ cm, expressing E in V/cm,

$$\Phi_E/\Phi_D = 3 \cdot 10^{-4} E. \qquad (12.15)$$

For particulates of smaller radius and for charged particulates, this ratio is even smaller. The rate of "drift" charging of the HDA particulates is thus comparable with the rate of diffusion charging only in fairly strong fields (above 3000 V/cm), e.g., in electrostatic filters, whereas in the atmosphere it is ignorable.

Passing to the experimental data on the charging kinetics of aerosol particulates by gas ions, we note that most of the measurements were carried out for comparatively large particulates at atmospheric pressure, i.e., at small Kn_i. Flanagan [216] measured the decrease in the steady-state concentration of gas ions in the air under the effect of a polydisperse aerosol generated by a hot nichrome wire, with a ranging from 0.01 to 0.13μ, i.e., for $Kn_i \gtrsim 1$. Assuming that the particulates acquired at most one elementary charge (which is true only for the lower part of this range), Flanagan determined from his measurements the sum of the fluxes Φ_0 and Φ_1', namely $\Phi_0 + \Phi_1' = 0.58 \, a$. This dependence for $Kn_i \gtrsim 1$ is not particularly consistent with the theory.

Liu, Whitby, and Yu [217] measured the charging kinetics of monodisperse dioctyl phthalate aerosols with $\bar{a} = 0.32$ and 0.67μ in an atmosphere with unipolar ions at pressures of 0.03–1 atm, i.e., for $Kn_i \lesssim 1$. They found that the charging of particulates as a function of time shows a good fit with Eq. (12.6), which does not contain Kn_i. The authors assumed that this equation was valid for any Kn_i. To justify this approach, they had to assume an ion mass 16 times the mass of neutral air molecules. As we have mentioned before, Eq. (12.6) is inapplicable for any Kn_i, and we will not even try to explain how the good fit was obtained. Note, however, that charging experiments carried out at low pressure provide an inherently inaccurate simulation of the process of charging of HDA particulates because of the image force effect, whose magnitude depends on particulate size, and not on Kn_i.

A number of authors measured the ratio of the content of charged and neutral aerosol particulates under steady-state conditions in a bipolar ion atmosphere and then calculated the ratio

$$\Phi_1'/\Phi_0 = 2N_0/N_1.$$

This work was carried out with polydisperse aerosols and can hardly be used as a check on theoretical calculations. The experiments of Pollak and Metnieks [218] with aerosols with $\bar{a} = (1–4) \cdot 10^{-6}$ cm generated by a hot nichrome wire showed that, in accordance with the theory, the experimental values of the ratio N_i/N_0 were higher than the results obtained from Boltzmann's equation for $\bar{a} < 1.35 \cdot 10^{-6}$ cm and lower for larger particulates. According to the authors, this effect is associated with the polydispersity of the aerosols, which introduces a distortion increasing with the increase in particulate size.

13. Thermophoresis and diffusiophoresis in HDA

In our analysis of these complex phenomena, we will only consider the range Kn $\geqq$ 1.

A particulate immersed in an inhomogeneous gaseous medium experiences a certain force, which is called *thermophoretic* if it is produced by a temperature gradient and *diffusiophoretic* if it is produced by a gradient of composition in the mixture. As in the drag problem, the force on the sur-surface S of the particulate is expressed by the integral

$$F_{Tk} = - \int dS \sum_{\pm} \int m_g[V_g V_g] f^{\pm} dV_g,$$

where f^+ and f^- are the distribution functions of the gas molecules with positive and negative projections of the velocity on the concentration gradient, V_g is the velocity of the gas molecules. The temperature or concentration gradient enters this equation implicitly, through the position dependent functions f^+ and f^-. The driving forces of the aerosol particulates in the temperature gradient are different for Kn $\gg$ 1 and Kn $\ll$ 1. For Kn $\gg$ 1, the phenomenon is entirely analogous to thermal diffusion. For Kn $\ll$ 1, the driving force is the temperature gradient inside the particulate and the tangential gas flow along its nonuniformly heated surface.

Unlike other transport phenomena, the thermophoretic force cannot be calculated using the zeroth approximation for f, not even in the region $\text{Kn} \gg 1$, since the gas is distinctly nonequilibrium.

The theory of thermal diffusion in a gas was applied to solve the problem of thermophoresis under free-molecular conditions by Einstein [219], Cawood [220], Clusius [221], and in more rigorous form by Deryagin and Bakanov [223], and also by Waldmann [222]. The last authors used the Chapman–Enskog theory in the Lorenz approximation, i.e., they assumed the mass of the aerosol particulates to be large compared to the mass of gas molecules and ignored the intermolecular collisions compared to the collisions of molecules with the particulate surface. Since a certain fraction of the molecules $(1 - \phi)$ are reflected specularly, and the remaining molecules suffer diffuse reflection, i.e., with a Maxwellian distribution corresponding to the temperature of the particulate surface, the thermophoretic force is independent of ϕ:

$$F_{Tk} = - (32/15)\,(a^2 \chi_g \nabla T / \bar{v}_g), \tag{13.1}$$

where ∇T is the temperature gradient, χ_g is the thermal conductivity of the gas. For a polyatomic gas, we should insert in (12.1) the translational component of thermal conductivity, $\chi_{gt} = 15\,k\eta/4m_g$, where η is the viscosity. Mason and Chapman [224] assumed elastic reflection for all the molecules; part of the molecules were reflected specularly, whereas the remainder (ϕ') were reflected in random directions, but without changing the magnitude of their velocity. Their result for F_{Tk} is greater by a factor of $1 + 4\pi/9$ than the figure obtained from (13.1). Monchik, Yun, and Mason [225], using the model of inelastic collisions developed by Wang Chang and Uhlenbeck [226], rigorously justified the applicability of Eq. (13.1) to polyatomic gases.

The theory of thermophoresis for intermediate Kn numbers was developed by Brock [227] who combined Bhatnagar's method with the Knudsen iteration. For a one-component polyatomic gas, the thermophoretic force according to Brock is

$$F_T = F_{Tk} \exp\{1 - [0.06 + 0.09\alpha_m + 0.28\,(1 - \alpha_m f a / 2\chi_g)\,]\,\text{Kn}\}, \tag{13.2}$$

where F_{Tk} is defined by (13.1), χ_g and χ_a are the thermal conductivities of the gas and the particulate, α_m is the coefficient of mechanical accommodation. This equation is equivalent to the empirical formula derived by

Schmitt [228]:

$$F_T = F_{Tk} \exp(-b/\text{Kn}),\qquad(13.3)$$

if we take for a monatomic gas

$$b = 0.06 + 0.37\alpha_m - 0.09\alpha_m\chi_a/2\chi_g,\qquad(13.4)$$

and for a polyatomic gas

$$b = 0.06 + 0.37\alpha_m - 0.28\alpha_m\alpha_T\chi_a/2\chi_g^*,\qquad(13.5)$$

where χ_g^* is the total thermal conductivity of a polyatomic gas, α_T is the thermal accommodation coefficient.

The values of the constant b from [229] are listed in Table 3.

Table 3

THE CONSTANT b IN EQ. (13.3) FROM VARIOUS
EXPERIMENTAL DATA

Source	Aerosol material	Gaseous atmosphere	b	α_m
[228]	Oil	Ar	0.39	0.8
[228]	Oil	N_2	0.38	0.81
[228]	Oil	CO_2	0.36	0.8
[228]	Oil	H_2	0.22	0.79
[240]	Tricresyl phosphate	Air	0.41	0.88
[241]	NaCl (amorphous)	Ar	0.47	0.91
[234]	NaCl (crystalline)	Air	0.46	0.91

Recently Dwyer [230] developed the theory of thermophoresis in the thirteen-moment approximation, which in his opinion is applicable to all Kn. This opinion, unfortunately, is unfounded, since the thirteen-moment approximation was developed for sliding conditions, i.e., Kn < 1 only. The application of this approximation to describe the coagulation kinetics [231] gave results which substantially departed from the exact theory already for Kn ≈ 1. For Kn → ∞ Dwyer's result does not coincide with the exact formula (13.1): it actually predicts a monotonic decrease of F_{Tk} with increasing Kn. Dwyer's conclusion that F_T may take on negative values for $0.9 < \alpha_m < 1.0$ is at variance with all the available experimental data [232].

Gardner [233] considered the thermophoresis of an evaporating particulate in its own rarefied vapor, without any carrier gas. He showed that the thermophoretic force in this case should be virtually the same as for a particulate without evaporation.

Experimental studies of the thermophoresis of silicone oil particulates in Ar, N_2, CO_2, and H_2, partly encompassing the range Kn > 1, were carried out by Schmitt [228]. He observed the displacement of an isolated droplet in a Millikan capacitor with the plates thermostated at various temperatures. The droplet radii ranged from 0.7 to 1.2μ, and the pressure was 10–760 mm Hg, so that the experiments covered the range of Kn numbers 0.05–3.3. For Kn $\approx$ 3.3, F_T is strictly proportional to a^2, which qualitatively fits the theory. For Kn > 0.2, the experimental data are approximated with Eq. (13.5), and for all gases, except H_2, $b = 0.38$. For Kn $\approx$ 3, the ratio of the theoretical to the experimental F_{Tk} values in Ar, N_2, CO_2, and H_2 is 1.05, 1.21, 1.22, and 1.38, respectively. For H_2, the experimental accuracy was the lowest. A similar method was applied by Schadt and Cadle [234] to sodium chloride and mercury particulates with radii of 0.22–1.15 and $0.096–1.145\mu$, respectively; these experiments covered the range 0.2 < Kn < 4.0. The measurements established that F_T is independent of the thermal conductivity of the particulates, i.e., they bear out the theoretical findings, although the actual result was closer to the theory of Cawood and Einstein, rather than to the calculations based on the Deryagin–Waldmann method.

In the analysis of diffusiophoresis, we should distinguish between two particular cases: 1) equimolecular counterdiffusion of two components, and 2) diffusion of one component through the other component at rest. The theory of diffusiophoresis for large Kn was developed by Waldmann [235], and also by Deryagin and Yalamov [236]. In case 1, the force on the particle is

$$F_{dk} = (8/3)\, a^2 n_g\, (2\pi kT)^{1/2} D_{12}\, (\nabla x_1)_\infty (m_{g2}^{1/2}\, (1 + \phi_2\pi/8) -$$
$$- m_{g1}^{1/2}\, (1 + \phi_1\pi/8)\,], \quad (13.6)$$

where D_{12} is the interdiffusion coefficient, x_1 is the mole fraction of component 1, $(\nabla x_1)_\infty$ is the concentration gradient of component 1 at infinity. For case 2,

$$F_{dk} = (8/3)\, a^2 n_g\, (2\pi m_{g1} kT)^{1/2} (1 + \phi_1\pi/8)\, D_{12}\, (\nabla x_1)_\infty /x_2. \quad (13.7)$$

Diffusiophoresis in a two-component mixture in the transitional range was investigated by Brock [237] using the Bhatnagar model, so that the application of the theory is limited to the case of molecules with near size and mass values. For case 1 and $x_1 \gg x_2$, taking $\phi_1 = \phi_2' = 1$, we get

$$F_d = F_{dk} \left\{ \frac{d_1^2 m_{g1}^{1/2}/d_{12}^2 - 0.311 \left[2m_{g1}m_{g2}/(m_{g1} + m_{g2}) \right]^{1/2}}{(m_{g2}^{1/2} - m_{g1}^{1/2})\,\text{Kn}} \right\}, \quad (13.8)$$

where d_1 and d_2 are the effective diameters of the molecules, $d_{12} = \frac{1}{2}(d_1 + d_2)$. In case 2 and $x_1 \ll x_2$,

$$F_d = F_{dk} \left\{ 1 - 0.071 \left[2m_{g2}/(m_{g2} + m_{g1}) \right]^{1/2} \text{Kn}^{-1} \right\}. \quad (13.9)$$

In Eqs. (13.8) and (13.9), F_{dk} is borrowed from (13.6) and (13.7), respectively.

Experimentally, diffusiophoresis for large and intermediate Kn was studied by Waldmann and Schmitt [238]. The measurements were carried out using silicone oil droplets at 20°C for various binary gas mixtures. For case 1, definite fit with Eq. (13.6) was observed in the mixtures CO_2/C_3H_8, N_2/C_2H_2, N_2/C_2H_4, N_2/C_2H_6, N_2/O_2, N_2/Ar. Brock [237] compared the calculated values of the ratio $v_d/D_{12}(\nabla x_1)_\infty$, where v_d is the velocity of displacement of a particulate in a composition gradient, with the results of Waldmann and Schmitt for $\text{Kn} \to \infty$; he obtained an adequate agreement for mixtures of nitrogen with O_2 and the above hydrocarbons. No fit between theory and experiment was observed for the N_2/Ar and CO_2/C_3H_8 mixtures, probably owing to the great disparity in the diameters of the components, which render inapplicable the simplified relaxation model of the rate equation of a binary mixture. For case 2, Brock obtained a satisfactory agreement for the asymptotic values of the ratio $v_d x_2/D_{12}(\nabla x_1)$ from theory and experiment for $\text{Kn} \to \infty$ in H_2O/N_2 mixtures.

In conclusion of this section on thermo- and diffusiophoresis, we can say that the fit between the theory in the transitional region and experiment is quite satisfactory, but a more general theory is required, based on a more strict treatment of the collision terms in the rate equations of a binary mixture.

Brock [239] pointed to still another form of motion of aerosol particles, *photodiffusiophoresis*, associated with changes in physical or chemical equilibrium between a particle and the gas medium under the action of electromagnetic radiation. Brock suggested the term *photothermophoresis* for the ordinary photophoresis, which is associated, as we know, with nonuniform heating of a particulate by incident radiation.

14. Coagulation of HDA

Smoluchowski has shown that the problem of Brownian coagulation is mathematically equivalent to the problem of vapor condensation on a particle. Therefore, the results of Sec. 9 concerning mass transport are fully applicable to coagulation. The only difference is that the Knudsen numbers in coagulation are defined somewhat differently: the mean free path of the gas molecules is replaced with the apparent (or effective, see p. 3) free path of the aerosol particulates, i.e., the average distance over which the particulate loses all correlation with the initial direction of its motion. This effective length is defined as $l_p = \bar{v}_p\tau$, where $\bar{v}_p = \sqrt{8kT/\pi m_p}$ is the thermal velocity of the particle, $\tau = m_p/B$, τ is the relaxation time, B is the mobility, m_p is the particle mass. In case of coagulation l_p should be further replaced with the appropriate parameter for the motion of two particulates, namely $l_p\sqrt{2}$. Finally, the radius of the absorbing sphere $r = 2a$ should be substituted for the particulate radius. Thus, the aerosol Knudsen number is expressed in the form $\text{Kn}_p = l_p\sqrt{2}/2a$. The values of l_p and Kn_p and spherical particles of unit density, suspended in air at NTP, are listed in Table 4.

Table 4

PARAMETERS CHARACTERIZING
THE COAGULATION OF HDA

Particle radius, Å	10	20	50	100	200	500	1000
l_p, Å	659	468	300	220	164	124	113
Kn_p	46.4	16.5	42.4	1.56	0.58	0.176	0.08
Coagulation constant, $K \cdot 10^{10}$ cm^3/sec:							
Smoluchowski	323	162	65.8	34.0	18.0	8.57	5.56
from Eq. (14.1)	4.42	9.80	14.0	19.8	19.8	31.3	
Sahni's theory	4.39	6.14	9.67	12.0	11.0	7.1	5.14

For large Kn_p, the coagulation constant should be expressed by the formula for the collision frequency between the gas molecules, i.e., the aerosol particulates may be treated as large gas molecules, so that the collision probability between them is not affected by the presence of another, lighter component (a Lorentz gas):

$$K_k = 2\sqrt{2}\pi a^2 \overline{V}_p = 4\,(3akT/\rho)^{1/2}, \tag{14.1}$$

where ρ is the particulate density. The applicability of this expression was rigorously justified by Hidy and Brock [242], who solved the rate equation for a Lorentz gas. Their expression for the coagulation constant allows for the electrostatic, interaction between the particulates, the effect of external fields, and the presence of velocity gradients; unfortunately, it does not include terms with molecular forces, which may prove to be quite significant in HDA coagulation, since the particulate size is comparable with the phonon wavelength. The molecular interaction potential for two spherical particulates of the same radius is described by Hamaker's equation

$$U(R) = \frac{A}{6}\left[\frac{2a^2}{R^2} + \frac{2a^2}{R^2 - 4a^2} + \ln\left(1 + \frac{4a^2}{R^2}\right)\right], \tag{14.2}$$

where R is the distance between the centers, $A = \pi q^2 \beta/6$, q is the number of molecules in unit volume, β is the London constant. From the equation of moments we readily find

$$b^2 = R_m^2\,[1 - U\,(R_m)/E], \tag{14.3}$$

where b is the impact parameter (see Figure 7), R_m is the minimum distance between the particulate centers, E is the total energy of the particulate. Calculations by successive approximations show that the function $b(R_m)$ for $2a < R_m < \infty$ should have a single minimum b_m. For $b < b_m$, R_m is imaginary, i.e., no minimum distance of the particle trajectory from the point O exists and the particles should collide. For $b > b_m$, R_m is real, and this corresponds to the case of no collision shown in Figure 6. For two moving particles, E is the kinetic energy of their relative motion for $V = 0$. The mean value of E is therefore $3kT$. In the absence of molecular forces $b_m = 2a$, i.e., the molecular forces increase collision cross section and, hence, the coagulation constant by a factor of $b_m^2/4a^2$.

Approximate solution of the rate equation enabled Brock [243] to determine the coagulation constant for intermediate Kn_p. In his derivation, each particulate is considered as a source of molecules with a perturbed distribution function. Colliding with the surface of another particulate, these molecules produce a repulsive force acting along the centerline.

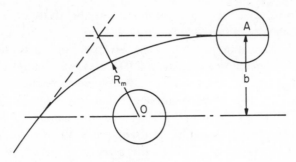

Figure 7
The trajectory of a particulate in a central attractive field.

Brock's expression is highly complex. For $Kn_p < 5$, Brock's correction term to the free-molecular relation (14.1) is close to unity; for $Kn_p \approx 2$ this factor slightly decreases and is a fraction of a percent less than unity; for $Kn_p < 2$, however, this correction factor rapidly increases as Kn_p decreases. Thus, for $Kn_p \approx 1$ Brock's equation predicts coagulation constants greater than the free-molecular values, i.e., the results contradict the underlying physical meaning of the formula.

The most exact data on the coagulation constant are apparently obtained by the method described in Sec. 9, using Eqs. (9.15) and 9.16). The values of the coagulation constant calculated by this method are listed in Table 4. For comparison, we give the results obtained from Eq. (14.1) and from Smoluchowski's equation, which is still favorably regarded by some experimentalists working with HDA.

Astakhov [231] determined the coagulation constant by the so-called thirteen-moment approximation. His expression, however, gives correct results only for $Kn_p < 1$.

The effect of molecular forces on coagulation at intermediate Kn_p numbers is a highly complex problem, since the delay of molecular forces over distances comparable with the phonon wavelength should be taken into consideration. No theoretical expression has been proposed which would allow for molecular forces for $Kn_p \approx 1$.

A recent result of Friedlander and Wang [244] indicates that no self-sustained particulate size distribution is established during aerosol coagulation under conditions when the coagulation constant is dependent on the Knudsen number.

There are very few experimental studies of the HDA coagulation. A significant difficulty in these studies are the considerable losses of particulates on the walls and in the communicating lines, associated with the high diffusion coefficient of the particulates. The classical method of studying coagulation processes, which calls for a periodic determination of particulate concentration in successive samples, is therefore inapplicable to HDA. O'Connor [245] and Quon [246], who used this method, obtained most puzzling results: the coagulation constant was found to increase monotonically for intermediate Kn_p with the decrease of particulate size, and this increase was even faster than what follows from the Smoluchowski equation. O'Connor worked with atmospheric condensation nuclei of unknown shape. Quon [246] used aerosols from incomplete combustion of hydrocarbons. Although in view of the low number density the aerosols probably consisted of individual particulates, and not aggregates, the question of their shape and polydispersity was left unclarified. In [101], some qualitative information regarding the coagulation of HDA was obtained.

Fuchs and Sutugin [98] used the flowthrough method to investigate HDA coagulation. An aerosol with a known high number density was passed in a laminar flow through a wide tube, and the inlet to outlet concentration ratio was measured nephelometrically. Before nephelometry, the HDA particulates were bulked to about 0.5μ by condensation by dibutyl phthalate vapor in the KUST instrument [97]. Separate experiments were carried out to determine the diffusion losses of particulates in the tube. To this end, a highly diluted aerosol was passed through the tube. The initial number density was determined with an ultramicroscope after dilution in a special device and bulking of HDA particulates in KUST. The coagulation constants determined by this method for monodisperse NaCl aerosols with mean particle radii of 25 and 45 Å were $13.4 \cdot 10^{-10}$ and $15.1 \cdot 10^{-10}$ cm^3/sec. The error of measurements was 11%. The correction term $b_m^2/4a^2$ to the free-molecular expression for the coagulation rate was calculated (this is the correction associated with the effect of molecular forces). The London constant calculated by Mayer [247] from the UV absorption bands of NaCl gave a correction of 2.19. The corrected coagulation constants were thus $10.2 \cdot 10^{-10}$ cm^3/sec for $\bar{a} = 25$ Å and $13.4 \cdot 10^{-10}$ cm^3/sec for $\bar{a} = 45$ Å. The second figure shows a good fit with the experimental data. The divergence between theory and experiment may be attributed to a certain polydispersity of the aerosols and inaccurate determination of the London constant.

The same procedure was used in [248] to determine the coagulation constant of monodisperse dioctyl sebacate aerosols with $\bar{a} = 100\,\text{Å}$. Kn_p in these experiments was 1.54. The experimental value of the coagulation constant was $13.4 \cdot 10^{-10}$ cm^3/sec, i.e., somewhat higher than the calculated figure in Table 4. The difference in this case may certainly be attributed to experimental errors and the effect of molecular forces.

Stockham [13] calculated the coagulational growth of silver particles and found that his experimental data could be fitted with theoretical results assuming a coagulation constant greater by a factor of 2–10 than in the free-molecular expression. The divergence between theory and experiment increased with the increase in the concentration of the vapor from which the aerosol formed. This was probably due to the formation of branched chain aggregates during the coagulation of aerosols with high weight concentrations. At low concentrations, Stockham's electron-microphotographs show compact aggregates, and the coagulation constant under these conditions was about double the free-molecular value. This result matches the factor of molecular forces calculated in [6] for the coagulation of silver particles: it was found to be 2.21.

The above results reveal a satisfactory agreement between theory and experiment with regard to the entire topic of HDA coagulation.

The theory of coagulation of charged aerosols in the molecular and the transitional regions is mathematically equivalent to the theory of charging of aerosol particles. In the free-molecular range, the coagulation constant of charged aerosols can be calculated using the expressions of Hidy and Brock [242]. For intermediate Kn_p, the corresponding calculations were carried out only by the boundary sphere method [249].

15. Capillary effects and the structure of HDA particulates

In addition to specific features of transport phenomena and optical and electrical properties of HDA, their thermodynamic properties also show some unique aspects, such as the increase in the vapor pressure of small particulates, expressed by the well-known Kelvin equation

$$\ln (p/p_0) = 2M\sigma/\rho RTa, \tag{15.1}$$

where p_0 is the equilibrium vapor pressure above a plane surface, p is the vapor pressure above a droplet of radius a, σ is the surface tension, ρ is the particulate density. The Kelvin effect becomes noticeable starting with particulates of $0.1-0.3\mu$, and it rapidly increases as a decreases. This is one of the reasons for the inherent instability of HDA from substances whose vapor pressure at room temperature is not exceedingly low. An indirect experimental verification of (15.1) was carried out by La Mer et al. [95, 250] using sulfuric acid and dibutyl phthalate droplets. Recently, direct observations were made of the rate of evaporation of liquid Pb and Bi droplets and solid Au particles on a coal substrate under an electron microscope [251]. The particulate size in these experiments was 100–500 Å, and they also confirmed the validity of the Kelvin equation.

The increased vapor pressure over a convex surface is sometimes termed a *capillary effect of the first kind. Capillary effects of the second kind* include the dependence of surface tension on surface curvature and the dependence of the pressure inside the droplet on its size and density.

Tolman [252] was the first to show that the dependence $\sigma(a)$ follows logically from the Gibbs thermodynamics. He derived the integral equation

$$\ln(\sigma/\sigma_0) = \int_a^\infty \frac{(2\delta_0/a^2)[1 + (\delta_0/a) + (\delta_0/a\sqrt{3})^2]}{1 + (2\delta_0/a)[1 + (\delta_0/a) + (\delta_0/a\sqrt{3})^2]} \, da, \quad (15.2)$$

where δ_0 is a parameter of the order of the interatomic distance. To a first approximation

$$\sigma/\sigma_0 = 1 - 2\delta_0/a. \quad (15.3)$$

Here σ_0 is the surface tension of a plane surface. Further developments in the theory of surface tension of curved surfaces were published in [253, 254, 255], where the authors avoided the inaccuracy of Tolman's method, i.e., taking the pressure inside the droplet equal to the pressure in a macroscopic liquid volume. The pressure in these studies was calculated from various equations of state. The following integral equation was obtained for δ_0:

$$\int_{-\infty}^{+\infty} (R + \delta_0)[p_0 - p_{xx}(R)] \, dR = 0, \quad (15.4)$$

where R is the radial coordinate, p_{xx} is the component of the pressure tensor describing the forces acting on an element of a spherical surface at right angles to R. The dependence $\sigma(a)$ is of considerable importance in the theory of spontaneous condensation. However, Kirkwood and Buff [256] have shown that the decrease in the density of small particulates must also be taken into consideration. The radial distribution of density and the surface tension of particulates with sizes of 3–25 molecular diameters were calculated by Plesner [257] using the equation of state for rigid spheres with an attraction potential inversely proportional to the sixth power of the distance. Plesner's result turned out to be most remarkable: the surface tension of these particulates is independent of size, being of the order of magnitude of $\frac{1}{3}\sigma_0$. This indicates that in a certain range of particulate sizes — a fairly narrow range, since the experimental dependence $\sigma(a)$ was not observed up to curvature values of 10^{-6} cm^{-1} — σ will decrease rapidly. This is a most unlikely result. According to Plesner, even at the center of very small droplets the density is different from the macroscopic value.

Shcherbakov [258, 259] noted that for very small droplets, the excess free energy is associated with the entire bulk of the droplet, and is not restricted to surface effects only. He also found that a straightforward insertion of the function $\sigma(a)$ in the Gibbs–Kelvin equation does not give a correct result, and that a term containing $d\sigma/da$ should also be introduced in this equation. A similar idea had been previously expressed by Gibbs and Kelvin, proceeding from intuitive considerations. Shcherbakov considers surface tension as the excess free energy per unit surface area of the particulate. The dependence $\sigma(a)$ according to Shcherbakov has the form

$$\sigma = \sigma_0\left(1 - \frac{2}{ka} + \frac{2}{k^2 a^2}\right), \tag{15.5}$$

where $1/k \approx 2$–5 Å. The contribution from the third term increases as a decreases, and the function $\sigma(a)$ therefore has a minimum, increasing with further decrease of particulate size for very small a. This result is at variance with the data of [252–255] and with the recent findings of Bellemans [260], who generalized the Mayer theory of the equation of state to the case of a system with an interface of arbitrary curvature.

A monotonically increasing function $\sigma(a)$ is also predicted by the theory of surface tension developed by Cahn and Hilliard [261, 262] and by Hart [263]. They regard the free energy density as a function of the molecular concentration and the concentration gradient. According to this theory, at high supersaturations in dense vapors, the liquid phase droplets have a very diffuse surface, and their density may differ only slightly from the vapor density. Developing this theory, Hart [264] reached a highly important conclusion. In the Gibbs theory of surface tension, σ is invariant irrespective of the choice of the exact position of the interface in the transition region between the phases. This invariance, however, is not observed for a curved surface, and the calculated value of σ actively depends on the position of the interface. In this case, the position of the interface and the value of the surface tension is found from the condition $d\sigma/dR = 0$, where R is the radial coordinate. Hart established that under certain conditions the function $\sigma(R)$ does not have a minimum, i.e., the very concept of surface tension is meaningless.

Okuyama and Zung [187] studied theoretically the dependence of the evaporation coefficient on curvature; they showed that the existence of this dependence is associated with the work to be done in increasing the surface of the droplet when yet another molecule is added to it. Their result is

$$\alpha_c = \alpha_{c0} \exp\left(-\varepsilon_0/kT\right), \tag{15.6}$$

where α_{c0} is the condensation coefficient on a plane surface, ε_0 is the activation energy of accretion on the particulate surface at $0°K$. According to these authors, ε_0 is equal to the surface energy of the droplet divided by the number of molecules n in the droplet. Then

$$\alpha_c = \alpha_{c0} \exp\left(-4\pi a^2 \sigma/nkT\right). \tag{15.7}$$

A significant lowering of σ_c, calculated using σ_0, becomes noticeable starting with $a \leq 100\,\text{Å}$; for water at $0°C$, with $a \sim 1\,\text{Å}$, α_c is 10^{-8} of the initial value of α_{c0}. The dependence $\sigma(a)$ will of course greatly suppress, if not offset altogether, this effect.

The lowering of the melting point of fine particulates is described by the Thomson equation:

$$\Delta T_m/T_m = 2\sigma v_l/La, \tag{15.8}$$

where ΔT_m is the lowering of the melting point relative to the macroscopic melting point T_m; v_l is the molar volume of the liquid; L is the heat of melting. The lowering of the heat of melting is responsible for the ability of HDA particulates to remain for a long time in an amorphous state, as well as for their tendency to coalesce or sinter on contact. Numerous results [32,

36, 46] show that the particulates of condensational HDA are usually amorphous. This finding is consistent with Ostwald's step rule. It seems quite natural if we remember that the free energy of formation of liquid phase nuclei from a vapor is lower than the free energy of formation of crystalline nuclei. Crystallization of the aerosol precipitates of aluminum [32] and various metal oxides [36] requires heating or exposure to X-rays. In dry air, NaCl particles of about 10^{-7} cm remain in an amorphous state for a long time [45]. In moist air, on the other hand, the large particles recrystallize in a few hours, and the small particles are formed crystalline to start with.

The ability of highly dispersed particulates to coalesce or sinter on contact has a most pronounced effect on their coagulation. The Tamman point, at which solids begin to sinter, is equal for most bodies to 0.6 of the melting point on the absolute scale [265]. The dependence of the Tamman temperature on particulate size was theoretically established by Higuchi [266].

A marked lowering of the specific surface of a nickel aerosol precipitate with ~ 30 Å particle size was noted in a few hours at room temperature [38]. It is remarkable that the composition of the atmosphere in which the precipitate was held apparently did not affect the process. The coagulation of silver aerosols with particle size of 100 Å was accompanied by particle sintering and coalescence at around 120–150° [6], i.e., at temperatures much lower than the normal Tamman point. The lowering of this temperature by a few hundred degrees for HDA precipitates of a number of refractory materials was also noted in [37]. It is the ability of the particles to sinter in coagulation that limits the prospects of preparation of highly dispersed powders through an intermediate aerosol state [6].

Bibliography

1. AMELIN, A.G. *Teoreticheskie osnovy obrazovaniya tumana (Theory of Fog Condensation)*, Chapter 2.—Izdatel'stvo "Khimiya". 1966. [English translation by ISRAEL PROGRAM FOR SCIENTIFIC TRANSLATIONS, Jerusalem 1967.]
2. LA MER, V.—*Industr. and Engng. Chem.*, **44**, 1229. 1952.
3. COURTNEY, W. *Heterogeneous Combustion*, **15**, 677.—New York, Academic Press. 1964.
4. CHRISTIANSEN, J.—*Acta chem. scand.*, **5**, 676. 1951.
5. REED, S.—*J. Chem. Phys.*, **20**, 208. 1952.
6. SUTUGIN, A.G. and N.A. FUCHS.—*J. Colloid and Interface Sci.*, **27**, 216. 1968.
7. HIDY, G.—*J. Colloid Sci.*, **20**, 123. 1965.
8. HIDY, G. and D. LILLY.—*J. Colloid Sci.*, **20**, 867.˙1965.
9. FRIEDLANDER, S. and A. SWIFT.—*J. Colloid Sci.*, **19**, 621. 1964.
10. FRIEDLANDER, S. and C. WANG.—*J. Colloid and Interface Sci.*, **22**, 126. 1966.
11. BEECKMANS, J.—*Canad. J. Chem.*, **43**, 2312. 1965.
12. ROSINSKI, J. and J. SNOW.—*J. Meteorol.*, **18**, 736. 1961.
13. STOCKHAM, J.—*Microscope*, **15**, 102. 1966.
14. STORGYN, D. and J. HIRSCHFELDER.—*J. Chem. Phys.*, **31**, 1531. 1959.
15. GRIFFIN, J. and P. SHERMAN.—*AIAA Journal*, **3**, 1813. 1965.
16. HILL, P., H. WITTIG, and E. DEMETRY.—*J. Heat and Mass Transfer*, **85**, 303. 1963.
17. WEGENER, P.—*Phys. Fluids*, **7**, 352. 1964.
18. TUNITSKII, N.N.—*Zhurnal Fizicheskoi Khimii*, **15**, 1061. 1941.
19. CORNER, J.—*Proc. Roy. Soc.*, **211A**, 417. 1952.
20. OSWATITSCH, K.—*Z. angew. Math. und Mech.*, **22**, 1. 1942.
21. BUIKOV, M.V. and V.P. BAKHANOV.—*Kolloidnyi Zhurnal*, **29**, No. 6. 1967.
22. DUNHAM, S.—*J. rech. atm.*, **2**, 331. 1966.
23. TESNER, P.A.—*Disc. Faraday Soc.*, **30**, 70. 1960.
24. O'CONNOR, T. et al.—*Geofis. pura e appl.*, **42**, 102. 1959.
25. LODGE, J. and B. TUFTS.—*J. Colloid Sci.*, **10**, 256. 1955.

26. GOYER, G. and F. PIDGEON.—*J. Colloid Sci.*, **11**, 697. 1956.
27. MEGAW, W. and R. WIFFEN.—*Proc. 1st Nat. Conf. Aerosols, Prague, 1962*, p. 511. Lublice. 1965.
28. O'CONNOR, T. and A. RODDY.—*J. rech. atm.*, **2**, 239. 1966.
29. GOLDSMITH, P. and F. MAY.—*Nature*, **210**, 475. 1966.
30. GEN, M.YA., M.V. ZISKIN, and YU.I. PETROV.—*Doklady AN SSSR*, **127**, 366. 1966.
31. GEN, M.YA., I.V. EREMIN, and YU.I. PETROV—*Zhurnal Tekhnicheskoi Fiziki*, **23**, 1407. 1959.
32. GEN, M.YA. and YU.I. PETROV.—*Pribory i Tekhnika Eksperimenta*, **4**, 162. 1963.
33. KIMOTO, K., Y. KAMIJA, and M. NANOYAMA.—*Jap. J. Appl. Phys.*, **2**, 702. 1963.
34. TURKEVITCH, J. In:—*Fundamental Phenomena Mater. Sci.*, **3**, 195. New York, Plenum Press. 1966.
35. SUTUGIN, A.G. and N.A. FUCHS—*Zhurnal Prikladnoi Khimii*, **39**, No. 12, 587. 1968.
36. HARVEY, J. and H. MATTHEWS.—*Disc. Faraday Soc.*, **30**, 113. 1960.
37. HOLMGREN, J., J. GIBSON, and G. SHEER.—*J. Electrochem. Soc.*, **111**, 362. 1964.
38. SELOVER, T.—*A.I.Ch.E. Journal*, **10**, 79. 1964.
39. SOFRONOV, N.YA.—*Kolloidnyi Zhurnal*, **29**, 327. 1967.
40. FUCHS, N.A. and A.G. SUTUGIN.—*Kolloidnyi Zhurnal*, **25**, 487. 1963.
41. SPURNY, K. and V. HAMPL.—*Coll. Czech. Chem. Comm.*, **30**, 507, 1965.
42. SUTUGIN, A.G.—*Kolloidnyi Zhurnal*, **27**, 789. 1965.
43. LA MER, V.—*Air Pollution (Proc. US Techn. Conf. Air Pollution)*, p. 607. Los Angeles. 1952.
44. MATIEVIC, E. and M. KERKER.—*Disc. Faraday Soc.*, **30**, 178. 1960.
45. MATIEVIC, E. and W. ESPENSCHIED.—*J. Colloid. Sci.*, **18**, 91. 1963.
46. ESPENSCHIED, W., E. MATIEVIC, and M. KERKER.—*J. Phys. Chem.*, **68**, 2831. 1964.
47. YOUNG, O. and J. MORRISON.—*J. Scient. Instrum.*, **31**, 90. 1954.
48. THOMPSON, F., G. ROSE, and J. MORRISON.—*J. Scient. Instrum.*, **32**, 324. 1955.
49. CRAIG, A. and R. MCINTOSH.—*Canad. J. Chem.*, **30**, 448. 1952.
50. THOMANN, H.—*Medd. Flygtechnik Försöksanst.*, **106**, 80. 1964.
51. THOMANN, H.—*Phys. Fluids*, **9**, 897. 1966.
52. NORGEN, C.—In: *AIAA. Progr. Astronaut. Aeronaut.*, **9**, 407. New York, Academic Press. 1963.
53. COX, L.—*J. Spacecraft and Rockets*, **4**, 86. 1967.
54. KARIORIS, F. and B. FISH.—*J. Colloid Sci.*, **17**, 155. 1962.

55. KARIORIS, F. and G. WOYCI.—*Adv. X-Ray Anal.*, **7**, 240. 1964.
56. WERLE, D.—*Armour Research Foundation, Contract Af*-19, Rept. 1955.
57. WINKEL, G. and G. JANDER.—*Kolloid-Z.*, **63**, 5. 1933.
58. JANDER, G. and G. WINKEL.—*Kolloid-Z.*, **65**, 292. 1935.
59. CAWOOD, W. and R. WHYTHLAW-GREY.—*Trans. Faraday Soc.*, **32**, 1048. 1936.
60. PETRYANOV, I.V., M.V. TIKHOMIROV, and N.N. TUNITSKII.—*Zhurnal Fizicheskoi Khimii*, **15**, 841. 1941.
61. WARBURG, O. and E. NEGELEIN.—*Biochem Z.*, **204**, 495. 1929.
62. BARZYNSKI, H. and D. HUMMEL—*Z. phys. Chem. (Frankfurt)*, **38**, 103. 1963.
63. KOGAN, YA.I.—*Doklady AN SSSR*, **161**, 386. 1965.
64. MAZLOVSKII, A.A. and N.A. FUCHS.—*Kolloidnyi Zhurnal*, **29**, 121. 1967.
65. CAILLAT. R. and J. CUER.—*Bull. Soc. chim. France*, No. 17, 152. 1959.
66. CUER, J., J. ELSTON, and S. TEICHNER.—*Bull. Soc. chim. France*, No. 1, 81. 1961.
67. ZAKUTINSKII, V.L. and I.S. BLYAKHER.—*Trudy UNIKhIM*, No. 14, 121. 1967.
68. THOMAS, A.—*Combustion and Flame*, **6**, 46. 1962.
69. AMELIN, A.G.—*Kolloidnyi Zhurnal*, **29**, 16. 1967.
70. BOGDANOV, V.S.—*Zhurnal Fizicheskoĭ Khimii*, **34**, 1044. 1960.
71. FUCHS, N.A. and N. OSHMANN.—*Acta physicochim. URSS*, **3**, 61. 1935.
72. ORR, C. et al.—*J. Meteorol.*, **15**, 240. 1958.
73. MATTESON, M. and W. STOBER.—*J. Colloid and Interface Sci.*, **23**, 203. 1967.
74. LIU, B., K. WHITBY, and H. YU.—*J. rech. atm.*, **2**, 396. 1966.
75. STERN, S.—*J. Appl. Phys.*, **30**, 952. 1959.
76. COX, L.—*Astronautics*, **1**, 56. 1962.
77. DROSIN, V.—*J. Colloid. Sci.*, **10**, 158. 1955.
78. GORBOVSKII, K. *Schetchiki atmosfernykh yader kondensatsii (Counters of Atmospheric Condensation Nuclei)*.—Leningrad. 1956.
79. NOLAN, P. and L. POLLAK.—*Proc. Roy. Irish Acad.*, **51A**, 9. 1946.
80. POLLAK, L.—*Geofis. pura e appl.*, **21**, 75. 1952.
81. POLLAK, L. and J. DALY.—*Geofis. pura e appl.*, **36**, 27. 1967.
82. POLLAK, L. and J. DALY.—*Geofis. pura e appl.*, **41**, 211. 1958.
83. POLLAK, L. and T. MURPHY.—*Geofis. pura e appl.*, **25**, 44. 1953.
84. POLLAK, L. and T. O'CONNOR.—*Geofis. pura e appl.*, **39**, 321. 1955.
85. NOLAN, P. and J. SCOTT.—*Proc. Roy. Irish Acad.*, **64**, 37. 1964.
86. VERZAR, F. —*Geofis. pura e appl.*, **29**, 192. 1954.
87. RICH, T.—*Geofis. pura e appl.*, **50**, 46. 1961.
88. POLLAK, L. and A. METNIEKS.—*Geofis. pura e appl.*, **47**, 123. 1960.
89. SKALA, G.—*Analyt. Chem.*, **35**, 702. 1965.
90. VONNEGUT, B.—*Proc. 1st Symp. Air Pollution*. Los Angeles. 1949.

91. Hüll, W. and R. Mueleisen.—*Naturwiss.*, **41**, 301. 1954.

92. Twomey, S.—*Geofis. pura e appl.*, **29**, 192. 1954.

93. Wieland, W.—*Z. angew. Math. Mech.*, **7**, 428. 1956.

94. La Mer, V., E. Inn, and J. Wilson.—*J. Colloid Sci.*, **5**, 471. 1950.

95. La Mer, V.—*US At. Energy Comm.*, Rept NYO–512. 1952.

96. Vlasenko, G.Ya., B.V. Deryagin, et al.—In: *Issledovanie oblakov, osadkov, grozovogo elektrichestva*, p. 285. Gidrometeoizdat. 1957.

97. Kogan, Ya.I. and A.V. Burnasheva.—*Zhurnal Fizicheskoi Khimii*, **39**, 2630. 1959.

98. Fuchs, N.A. and A.G. Sutugin.—*Kolloidnyi Zhurnal*, **28**, 131. 1966.

99. Aitken, J.—*Proc. Roy. Soc.*, **A37**, 215. 1916.

100. Nolan, P. and L. Pollak.—*Geofis. pura e appl.*, **31**, 60. 1955.

101. Yaffe, Y. and R. Cadle.—*J. Chem. Phys.*, **62**, 510. 1958.

102. Twomey, S.—*J. rech. atm.*, **1**, 101. 1963.

103. Orr, C. and T. Wilson.—*J. Colloid Sci.*, **19**, 571. 1964.

104. Twomey, S. and G. Severince.—*J. rech. atm.*, **1**, 81. 1961.

105. Buikov, M.V.—*Kolloidnyi Zhurnal*, **28**, 164. 1966.

106. Buikov, M.V.—*Kolloidnyi Zhurnal*, **28**, 636. 1966.

107. Siksna, R.—*Geofis. pura e appl.*, **50**, 23. 1961.

108. Vonnegut, B. and R. Neubauer.—*Bull. Amer. Meteorol. Soc.*, **34**, 163. 1953.

109. Soudain, G.—*J. sci. de la meteorol.*, **3**, 137. 1951.

110. Binek, M.—*Rev. trimestr. Ass. prevent. pollution atm.*, **5**, 91. 1963.

111. Langevin, P.—*J. Phys.*, **4**, 322. 1965.

112. Lisovskii, P.V.—*Acta physicochim. URSS*, **13**, 157. 1940.

113. Israel, H.—*Quart. J. Roy. Met. Soc.*, **35**, 341. 1932.

114. McGreevy, G.—*Arch. Meteorol. Geophys. und Bioklimatol.*, **A14**, 318. 1964.

115. Keefe, D., P. Nolan, and J. Scott.—*Proc. Roy. Irish Acad.*, **66A**, 17. 1968.

116. Wright, H.—*Proc. Phys. Soc.*, **48**, 675. 1936.

117. Rich, T.—*Int. J. Air. Wat. Pollution*, **1**, 288. 1959.

118. Rich, T., A. Metnieks, and L. Pollak.—*Geofis. pura e appl.*, **44**, 233. 1959.

119. Fuchs, N.A.—*Geofis. pura e appl.*, **56**, 185. 1963.

120. Nolan, P. and V. Guerrini.—*Proc. Roy. Irish Acad.*, **43**, 5. 1935.

121. Radushkevich, L.V.—*Acta physicochim. URSS*, **11**, 265. 1939.

122. Gormley, P. and M. Kennedy.—*Proc. Roy. Irish Acad.*, **52A**, 166. 1949.

123. Gormley, P. and M. Kennedy.—*Proc. Roy. Irish Acad.*, **45A**, 59. 1938.

124. DeMarcus, W. and J. Thomas.—*US At. Energy Comm.*, Rept. ORNL–1413. 1952.

125. Twomey, S.—*J. Franklin Inst.*, **275**, 1969. 1963.

126. Rodebush, W.—*US At. Energy Comm.*, Rept. OSRD–2050. 1943.

127. Pollak, L. and A. Metnieks.—*Geofis. pura e appl.*, **37**, 183. 1957.

128. POLLAK, L. and A. METNIEKS.—*Geofis. pura e appl.*, **44**, 224. 1959.
129. POLLAK, L., J. DALY, and A. METNIEKS.—*Geofis, pura e appl.*, **45**, 249. 1960.
130. NOLAN, P. and J. SCOTT.—*Proc. Roy. Irish Acad.*, **65A**, 39. 1965.
131. FUCHS, N.A., I.B. STECHKINA, and V.A. STAROSEL'SKII.—*Inzhenerno-Fizicheskii Zhurnal*, **5**, 100. 1962.
132. TWOMEY, S.—*J. Franklin Inst.*, **275**, 163. 1963.
133. METNIEKS, A.—*Pure Appl. Geoph.*, **61**, 183. 1965.
134. THOMAS, J.—*J. Colloid Sci.*, **10**, 46. 1955.
135. THOMAS, J.—*J. Colloid Sci.*, **11**, 107. 1956.
136. TWOMEY, S. and G. SEVERINSE.—*J. Atm. Sci.*, **20**, 392. 1963.
137. KOGAN, YA.I. and R. REPINA.—*Doklady AN SSSR*, **111**, 851. 1956.
138. BRICARD, J. et al.—*Staub-Reinhalt. Luft*, **24**, 345. 1964.
139. CHAMBERLAINE, A. and W. MEGAW.—*Geofis. pura e appl.*, **36**, 273. 1957.
140. KORPUSOV, V., V. KIRICHENKO, and B. OGORODNIKOV.—*Atomnaya Energiya*, **17**, 221. 1964.
141. BEREZHNOI, V.—*Atomnaya Energiya*, **18**, 342. 1965.
142. FUCHS, N.A. *Uspekhi mekhaniki aerozolei (Advances in the Mechanics of Aerosols)*, p. 77.—Izdatel'stvo AN SSSR. 1961.
143. ANDERSON, H., P. NOLAN, and T. O'CONNOR.—*Proc. Roy. Irish Acad.*, **66A**, 69. 1968.
144. MEGAW, W. and K. WIFFEN.—*J. rech. atm.*, **1**, 173. 1963.
145. NOLAN, P. and P. KENNY.—*J. Atm. Terrestrial Phys.*, **3**, 181. 1953.
146. KIRSCH, A.A. and N.A. FUCHS.—*Kolloidnyi Zhurnal*, **30**, 285. 1968.
147. KIRSCH, A. and N. FUCHS.—*Ann. Occup. Hygiene*, **11**, 642. 1968.
148. RICH, T.—*J. rech. atm.*, **2**, 79. 1966.
149. POLLAK, L., T. O'CONNOR, and A. METNIEKS.—*Geofis. pure e appl.*, **34**, 177. 1956.
150. POLLAK, L. and T. MURPHY.—*Geofis. pura e appl.*, **31**, 66. 1955.
151. LABEIRIE, J.—*Brit. J. Appl. Phys.*, Suppl. No. 3, *Discussion*. 1954.
152. FUCHS, N.A. and S.YA. YANKOVSKII.—*Kolloidnyi Zhurnal*, **21**, 133. 1959.
153. ORR, C. and T. WILSON.—*J. Colloid. Sci.*, **19**, 571. 1964.
154. ORR, C. and R. MARTIN.—*Rev. Scient. Instrum.*, **29**, 129. 1958.
155. LANGER, G., J. PIERRARD, and G. YAMATE.—*Int. J. Air Water Pollut.*, **8**, 167. 1964.
156. BAKER, R., Editor.—*Adv. Optical and Electron Microscopy.*—New York, Academic Press. 1966.
157. CADLE, R. *Particle Size Determination.*—New York. Interscience Publ. 1955.
158. MEGAW, W. and R. WIFFEN.—*Int. J. Air Water. Pollut.*, **7**, 501. 1963.
159. KALMUS, E.—*J. Appl. Phys.*, **25**, 87. 1954.
160. BIGG, E., G. MILES, and K. HEFFERMANN.—*J. Meteorol.*, **18**, 804. 1961.
161. MOSSOP, S. and H. THORNDIKE.—*J. Appl. Math. Phys.*, **5**, 474. 1966.

162. CARTWRIGHT, J.—*Quart. J. Roy. Meteorol. Soc.*, **82**, 82. 1956.

163. STECHKINA, I.B.—*Proceedings of the Third Inter-University Conference on the Problems of Stability and Dynamics of Disperse Systems*, p. 47. Odessa, Izdatel'stvo Odesskogo Gosudarstvennogo Universiteta. 1968.

164. LEKHTMAKHER, S.O., L.S. RUZER, et al.—*Ibid.*, p. 39.

165. LORD, R. and A. HARBOUR.—*AIAA Journal*, **6**, 2. 1968.

166. FUCHS, N.A. *Rost i isparenie kapel' v gazoobraznoi srede (Growth and Evaporation of Drops in Gaseous Media)*, sec. 14 and 17.—Moscow, Izdatel'stvo AN SSSR. 1958.

167. JER RU MAU.—*Ind. Eng. Chem. Fundamentals*, **6**, 504. 1967.

168. SCHÄFER, K., W. RATING, and A. ENSKOG.—*Ann. Phys.*, **42**, 176. 1942.

169. FRISH, H. and F. COLLINS.—*J. Chem. Phys.*, **20**, 797. 1952.

170. TIMOFEEV, M. and M. SHVETS.—*Meteorologiya i Gidrologiya*, No. 2, 9. 1968.

171. SHERMAN, F.—*Rarefied Gas Dynamics, 3rd Symp.*, **2**, 228. New York, Academic Press. 1963.

172. WELLANDER, P.—*Arkiv fysik*, **7**, 507. .1954.

173. BHATNAGAR, P., E. GROSS, and H. KROOK.—*Phys. Rev.*, **94**, 511. 1954.

174. BROCK, J.—*J. Colloid and Interface Sci.*, **22**, 513. 1966.

175. BROCK, J.—*J. Colloid and Interface Sci.*, **24**, 344. 1967.

176. JEANS, J. *Dynamic Theory of Gases*, 4th Ed., Chapt. 13.—New York, Dover Publ. Inc. 1925.

177. DAVISON, B. *Theory of Neutron Transport.*—Oxford Clarendon Press. 1957.

178. SAHNI, D.—*J. Nucl. Energy*, **20**, 915. 1966.

179. SMIRNOV, V.I.—*Trudy TsAO*, No. 55, 86. 1964.

180. BRADLEY, R. et al.—*Proc. Roy. Soc.*, **A186**, 368. 1948; **A198**, 226, 238. 1949.

181. MONCHIK, L. and H. REISS.—*J. Chem. Phys.*, **22**, 831. 1954.

182. BROCK, J.—*J. Colloid and Interface Sci.*, **24**, 344. 1967.

183. BROCK, J.—*J. Phys. Chem.*, **68**, 2857. 1964.

184. SMIRNOV, V.I.—*Trudy TsAO*, No. 92, sec. 3. 1968.

185. SANG-WOOK, KANG.—*AIAA Journal*, **5**, 1288. 1967.

186. MERRIT, C. and R. WEATHERSTONE.—*AIAA Journal*, **5**, 190. 1967.

187. OKUYAMA, M. and J. ZUNG.—*J. Chem. Phys.*, **46**, 1580. 1967.

188. HORN, K., V. BEVC, and M. KAPLAN.—*AIAA Journal*, **5**, 721. 1967.

189. CHAMBERLAINE, A. et al.—*Disc. Faraday Soc.*, **30**, 162. 1960.

190. BRIGGS, G.H.—*Philos. Mag.*, **50**, 630. 1928.

191. MILLIKAN, R.—*Phys. Rev.*, **21**, No. 217. 1923; **22**, 1. 1923.

192. EPSTEIN, P.—*Phys. Rev.*, **23**, 710. 1924.

193. FUCHS, N.A. and I.B. STECHKINA.—*Zhurnal Tekhnicheskoi Fiziki*, **33**, 132. 1963.

194. LIU VI CHANG, PANG SING CHEN, and H. JEW.—*Phys. Fluids*, **8**, 788. 1968.

195. WILLIS, D.—*Phys. Fluids*, **9**, 2522. 1966.

196. CERCIGNANI, C. and C. PAGANI.—*Phys. Fluids*, **9**, 1167. 1966.
197. CERCIGNANI, C. and C. PAGANI.—*Phys. Fluids*, **11**, 1395, 1399. 1968.
198. BROCK, J.—*J. Phys. Chem.*, **68**, 2863. 1964.
199. KENNARD, E. *Kinetic Theory of Gases*, p. 313.—N.Y., McGraw–Hill. 1938.
200. SPRINGER, G. and S. TSAI.—*Phys. Fluids*, **8**, 1361. 1965.
201. LEES, L.—*J. Soc. Ind. Appl. Math.*, **13**, 278. 1965.
202. BROCK, J.—*Phys. Fluids*, **9**, 1601. 1966.
203. CERCIGNANI, C. and C. PAGANI.—*Rarefied Gas Dynamics, 5th Symp.*, **1**, 555. New York, Academic Press. 1967.
204. TAKAO, H.—*Rarefied Gas Dynamics, 3rd Symp.*, **2**, 102. Academic Press. 1963.
205. HARBOUR, P.—Ph.D. Thesis, Cambridge. 1963.
206. JUNGE, C. *Air Chemistry and Radioactivity.*—Academic Press. 1963.
207. GENTRY, G. and J. BROCK.—*J. Chem. Phys.*, **47**, 64. 1967.
208. LIU B., K. WHITBY, and H. YU.—*J. Colloid Sci.*, **23**, 367. 1967.
209. KEEFE, D. and P. NOLAN.—*Proc. Roy. Irish Acad.*, **62A**, 8. 1962.
210. KEEFE, D., P. NOLAN, and T. RICH.—*Proc. Roy. Irish Acad. Sci.*, **60A**, 27. 1959.
211. KEEFE, D., P. NOLAN, and J. SCOTT.—*Proc. Irish Acad.*, **66A**, 17. 1968.
212. BRICARD, J. *Problem of Atmospheric and Space Electricity*, p. 82.—Amsterdam, Elsevier. 1965.
213. SIKSNA, R.—*J. rech. atm.*, **1**, 137. 1963.
214. FUCHS, N.A.—*Geofis. pura e appl.*, **56**, 185. 1963.
215. NATANSON, G. L.—*Zhurnal Tekhnicheskoi Fiziki*, **30**, 573. 1960.
216. FLANAGAN, V.—*Pure Appl. Geoph.*, **64**, 197. 1966.
217. LIU, B., K. WHITBY, and H. YU.—*Brit. J. Appl. Phys.*, **18**. 1967.
218. POLLAK, L. and A. METNIEKS.—*Geofis. pura e appl.*, **51**, 225. 1962.
219. EINSTEIN, A.—*Z. Phys.*, **27**, 1. 1924.
220. CAWOOD, W.—*Trans. Faraday Soc.*, **32**, 1068. 1936.
221. CLUSIUS, K.—*Z. VDI, Verfahrenstechnik*, No. 2, 23. 1941.
222. WALDMANN, L.—*Z. Naturforsch.*, **14A**, 589. 1959.
223. DERYAGIN, B.V. and S.P. BAKANOV.—*Kolloidnyi Zhurnal*, **21**, 377. 1959.
224. MASON, E. and S. CHAPMAN.—*J. Chem. Phys.*, **36**, 627. 1963.
225. MONCHIK, L., K. YUN, and E. MASON.—*J. Chem. Phys.*, **39**, 654. 1963.
226. WANG CHANG and G. UHLENBECK.—*Eng. Res. Inst. Univ. Michigan, Rept.* 1951.
227. BROCK, J.—*J. Colloid and Interface Sci.*, **23**, 448. 1967.
228. SCHMITT, K.—*Z. Naturforsch.*, **14A**, 870. 1959.
229. BROCK, J.—*J. Colloid and Interface Sci.*, **25**, 392. 1967.
230. DWYER, H.—*Phys. Fluids*, **10**, 976. 1967.
231. ASTAKHOV, A.V.—*Doklady AN SSSR*, **161**, 1114. 1965.
232. BROCK, J.—*Phys. Fluids*, **11**, 922. 1968.

233. GARDNER, G.—*Chem. Engng. Sci.*, **23**, 29. 1968.
234. SCHADT, G. and R. CADLE.—*J. Colloid Sci.*, **12**, 356. 1957.
235. WALDMANN, L.—*Rarefied Gas Dynamics*, p. 323, N. TALBOT, editor. New York, Academic Press. 1960.
236. DERYAGIN, B.V. and YU.I. YALAMOV.—*J. Colloid and Interface Sci.*, **22**, 117. 1966.
237. BROCK, J.—*J. Colloid and Interface Sci.*, **27**, 95. 1968.
238. WALDMANN, L. and K. SCHMITT.—*Z. Naturforsch.*, **15A**, 843. 1960.
239. BROCK, J.—*J. Phys. Chem.*, **72**, 747. 1968.
240. ROSENBLATT, P. and V. LAMER.—*Phys. Rev.*, **70**, 385. 1946.
241. JACOBSEN, S. and J. BROCK.—*J. Colloid Sci.*, **20**, 544. 1965.
242. HIDY, G. and J. BROCK.—*J. Appl. Phys.*, **36**, 1857. 1965.
243. HIDY, G. and J. BROCK.—*J. Colloid Sci.*, **20**, 477. 1965.
244. FRIEDLANDER, S. and C. WANG.—*J. Colloid and Interface Sci.*, **24**, 170. 1967.
245. O'CONNOR, T.—*Geofis, pura e appl.*, **31**, 107. 1955.
246. QUON, J.—*Int. J. Air Water Pollut.*, **8**, 355. 1964.
247. MAYER, J.—*J. Chem. Phys.*, **1**, 270. 1933.
248. SUTUGIN, A.G.—*Kolloidnyi Zhurnal*, **29**, 852. 1967.
249. ZEBEL, L.—In: *Aerosol Science*, N. DAVIES, editor.—London, Academic Press. 1966.
250. LA MER, V. and R. GRUEN.—*Trans. Faraday Soc.*, **48**, 410. 1952.
251. BLACKMAN, M., N. LISGARTEN, and L. SKINNER.—*Nature*, **217**, 1245. 1967.
252. TOLMAN, R.—*J. Chem. Phys.*, **17**, 33. 1949.
253. BUFF, F.—*J. Chem. Phys.*, **19**, 159. 1951.
254. BUFF, F.—*Disc. Faraday Soc.*, **30**, 52. 1960.
     HILL, T.—*J. Chem. Phys.*, **56**, 527. 1952.
256. KIRKWOOD, B. and F. BUFF.—*J. Chem. Phys.*, **18**, 991. 1950.
257. PLESNER, I.—*J. Chem. Phys.*, **40**, 1510. 1964.
258. SHCHERBAKOV, L.M.—*Kolloidnyi Zhurnal*, **14**, 379. 1952.
259. SHCHERBAKOV, L.M., P.P. RYAZANTSEV, and N.P. FILIPPOV.—*Kolloidnyi Zhurnal*, **23**, 338. 1961.
260. BELLEMANS, A.—*Physica*, **29**, 548. 1963.
261. CAHN, J. and J. HILLIARD.—*J. Chem. Phys.*, **28**, 258. 1958.
262. CAHN, J.—*J. Chem. Phys.*, **31**, 688. 1959.
263. HART, E.—*Phys. Rev.*, **113**, 412. 1959.
264. HART, E.—In: *Fundamental Phenomena Material Sci.*, p. 37. New York, Plenum Press. 1966.
265. TAMMAN, L.—*Z. anorgan. allgem. Chem.*, **110**, 166. 1928.
266. HIGUCHI, J.—*Science Rept. Tôhoku Univ.*, 1st Ser., **33**, 237. 1949.

Name Index

Numbers in *italics* indicate references to pages in which the authors' names are not mentioned in the text

Subject Index

Page numbers in *italic* type indicate figures; page numbers followed by (T) indicate tables

101

Printed in Israel
Manufactured at the Israel Program for Scientific Translations, Jerusalem